N.C. Chauhan, M.V. Kartikeyan, and A. Mittal

Soft Computing Methods for Microwave and Millimeter-Wave Design Problems

T0180109

Studies in Computational Intelligence, Volume 392

Editor-in-Chief

Prof. Janusz Kacprzyk
Systems Research Institute
Polish Academy of Sciences
ul. Newelska 6
01-447 Warsaw
Poland
E-mail: kacprzyk@ibspan.waw.pl

Further volumes of this series can be found on our
homepage: springer.com

Vol. 371. Leonid Perlovsky, Ross Deming, and
Roman Ilin (Eds.)
*Emotional Cognitive Neural Algorithms with Engineering
Applications*, 2011
ISBN 978-3-642-22829-2

Vol. 372. António E. Ruano and
Annamária R. Várkonyi-Kóczy (Eds.)
New Advances in Intelligent Signal Processing, 2011
ISBN 978-3-642-11738-1

Vol. 373. Oleg Okun, Giorgio Valentini, and Matteo Re (Eds.)
Ensembles in Machine Learning Applications, 2011
ISBN 978-3-642-22909-1

Vol. 374. Dimitri Plemenos and Georgios Miaoulis (Eds.)
Intelligent Computer Graphics 2011, 2011
ISBN 978-3-642-22906-0

Vol. 375. Marenglen Biba and Fatos Xhafa (Eds.)
Learning Structure and Schemas from Documents, 2011
ISBN 978-3-642-22912-1

Vol. 376. Toyohide Watanabe and Lakhmi C. Jain (Eds.)
Innovations in Intelligent Machines – 2, 2011
ISBN 978-3-642-23189-6

Vol. 377. Roger Lee (Ed.)
*Software Engineering Research, Management
and Applications 2011*, 2011
ISBN 978-3-642-23201-5

Vol. 378. János Fodor, Ryszard Klempous, and
Carmen Paz Suárez Araujo (Eds.)
Recent Advances in Intelligent Engineering Systems, 2011
ISBN 978-3-642-23228-2

Vol. 379. Ferrante Neri, Carlos Cotta, and
Pablo Moscato (Eds.)
Handbook of Memetic Algorithms, 2011
ISBN 978-3-642-23246-6

Vol. 380. Anthony Brabazon, Michael O'Neill, and
Dietmar Maringer (Eds.)
Natural Computing in Computational Finance, 2011
ISBN 978-3-642-23335-7

Vol. 381. Radosław Katarzyniak, Tzu-Fu Chiu,
Chao-Fu Hong, and Ngoc Thanh Nguyen (Eds.)
*Semantic Methods for Knowledge Management and
Communication*, 2011
ISBN 978-3-642-23417-0

Vol. 382. F.M.T. Brazier, Kees Nieuwenhuis, Gregor Pavlin,
Martijn Warnier, and Costin Badica (Eds.)
Intelligent Distributed Computing V, 2011
ISBN 978-3-642-24012-6

Vol. 383. Takayuki Ito, Minjie Zhang, Valentin Robu,
Shaheen Fatima, and Tokuro Matsuo (Eds.)
*New Trends in Agent-Based Complex Automated
Negotiations*, 2012
ISBN 978-3-642-24695-1

Vol. 384. Daphna Weinshall, Jörn Anemüller,
and Luc van Gool (Eds.)
Detection and Identification of Rare Audiovisual Cues, 2012
ISBN 978-3-642-24033-1

Vol. 385. Alex Graves
*Supervised Sequence Labelling with Recurrent Neural
Networks*, 2012
ISBN 978-3-642-24796-5

Vol. 386. Marek R. Ogiela and Lakhmi C. Jain (Eds.)
*Computational Intelligence Paradigms in Advanced Pattern
Classification*, 2012
ISBN 978-3-642-24048-5

Vol. 387. David Alejandro Pelta, Natalio Krasnogor,
Dan Dumitrescu, Camelia Chira, and Rodica Lung (Eds.)
*Nature Inspired Cooperative Strategies for Optimization
(NICSO 2011)*, 2011
ISBN 978-3-642-24093-5

Vol. 388. Tiansi Dong
Recognizing Variable Environments, 2012
ISBN 978-3-642-24057-7

Vol. 389. Patricia Melin
*Modular Neural Networks and Type-2 Fuzzy Systems for
Pattern Recognition*, 2012
ISBN 978-3-642-24138-3

Vol. 390. Robert Bembenik, Lukasz Skonieczny,
Henryk Rybiński, and Marek Niezgódka (Eds.)
*Intelligent Tools for Building a Scientific Information
Platform*, 2012
ISBN 978-3-642-24808-5

Vol. 391. Herwig Unger, Kyandoghere Kyamaky,
and Janusz Kacprzyk (Eds.)
Autonomous Systems: Developments and Trends, 2012
ISBN 978-3-642-24805-4

Vol. 392. N.C. Chauhan, M.V. Kartikeyan,
and A. Mittal
*Soft Computing Methods for Microwave and Millimeter-Wave
Design Problems*, 2012
ISBN 978-3-642-25562-5

N.C. Chauhan, M.V. Kartikeyan, and A. Mittal

Soft Computing Methods for Microwave and Millimeter-Wave Design Problems

 Springer

Authors

Dr. N.C. Chauhan
Department of Information Technology
A.D. Patel Institute of Technology
Anand 388 121
India
E-mail: narendracchauhan@gmail.com

Prof. Dr. A. Mittal
Director (Research) Graphic Era University
Dehradun 248 002
India
E-mail: dr.ankush.mittal@gmail.com

Prof. Dr. M.V. Kartikeyan
Indian Institute of Technology Roorkee
Department of Electronics and Computer
   Engineering
Millimeter Wave Laboratory
Roorkee 247 667
India
E-mail: kartik@iitr.ernet.in

ISSN 1860-949X
ISBN 978-3-642-43760-1
DOI 10.1007/978-3-642-25563-2
Springer Heidelberg New York Dordrecht London

e-ISSN 1860-9503
ISBN 978-3-642-25563-2 (eBook)

Foreword

It is my great pleasure to write this Foreword for the book entitled: "*Soft Computing Methods for Microwave and Millimeter Wave Design Problems*" of Professor Kartikeyan Machavaram and his colleagues Associate Professor Narendra Chauhan and Professor Ankush Mittal. After a careful review of the manuscript I can confirm that the book's title justly represents its contents, motivation and scope.

Soft computing methods have opened a wonder world for the design optimization and related problems in engineering disciplines. Classical iterative methods or equivalently parametric analysis procedures have duly been overtaken by an unorthodox swarm of progressions that follow nature and its evolutionary phenomenon to mimic or replicate processes to be optimized. The RF community has just begun to exploit the advantages offered by the soft computing methods for enormous design problems. The onus is on two important aspects: the first one is design perfection and the second is computational economy. Microwave and millimeter wave design problems are to be dealt precariously as they need a special attention to carry out the design and optimization cycles of the devices, components, and systems to their perfection.

The book has been presented in eight chapters. Design challenges and objectives are outlined in Chapter-1. Chapter-2 deals with an overview of soft computing methods. A comprehensive review of soft computing techniques in the perspective of microwave and millimeter wave design problems has been given in Chapter-3. A variety of diversified design problems in the microwave/millimeter wave domain are presented in Chapters 4-7. Chapter-8 gives a brief account of summary and outlook.

This book treats the subject to meet its objectives and will serve the needs of scientists, engineers and designers in RF & Microwave Engineering. As a whole, this book presents a close perspective of a new methodology for the design and optimization of complex problems in the microwave/millimeter wave domain. I congratulate the authors for coming up with an excellent book embedded with their own contributions for the benefit of the RF and microwave community and I look forward that these relentless efforts will be received well by the readers.

Karlsruhe, 25 September 2011   M.K. Thumm

Preface

The growing commercial market of Microwave/ Millimeter wave industry over the past decade has led to the explosion of interests and opportunities for the design and development of microwave components. Particularly, for the communications industry, the onus is on the successful development of components in shortest possible time with optimized cost. There is also a class of critical design applications such as design of high power devices and components where the accuracy requirement is the prime goal. The design of the most microwave components requires the use of commercially available electromagnetic (EM) simulation tools for their analysis. In the design process, the simulations are carried out by varying the design parameters until the desired response is obtained. The optimization of design parameters by manual searching is a cumbersome and time consuming process, and the chances to get local minima are very high. Moreover, increasing number of design parameters or widening the search range makes it difficult to converge to the global optima.

Soft computing methods such as Genetic Algorithm (GA), Artificial Neural Network (ANN) and Fuzzy Logic (FL) have been widely used by EM researchers for microwave design since last decade [6, 3, 7]. Soft computing methods play important role in the design and optimization of many engineering disciplines including microwave domain. The aim of these methods is to tolerate imprecision, uncertainty, and approximation to achieve robust and low cost solution in a small time frame [3]. However, these methods suffer from certain drawbacks. GA is a powerful optimization tool, but it requires a large number of iterations to achieve convergence and to arrive at an optimum solution. ANN has proved its efficiency in modeling microwave components, but it has also suffered with generalization problems. Moreover, due to very small wavelengths involved in microwave design, which requires high precision, it is not easy to model components using conventional methods. In all, modeling and optimization are essential parts and powerful tools for the microwave/millimeter wave design task but they must be applied judiciously.

Our present work deals with the development and use of soft computing based methods for tackling challenging design problems in microwave/millimeter wave domain. Our aim in the development and investigation of these methods is to obtain the designs in small time frame while improving the accuracy of the design for a wide range of applications. In order to achieve this goal, a few diverse design problems of microwave field, representing varied challenges in the design, such as different microstrip antennas, microwave filters, a microstrip–via, and also some critical high power components such as nonlinear tapers and RF–windows have been considered as case-study design problems. Different design methodologies are developed for these applications.

The book is organized in eight chapters with adequate space dedicated for the soft computing methods, their review for microwave/millimeter wave design problems followed by specific case–study problems to infuse better insight and understanding of the subject. Finally, summary and suggestions are given for the future scope.

During the course of the preparation of this book, our colleagues at Indian Institute of Technology Roorkee and at A D Patel Institute of Technology helped us immensely. We sincerely thank Professor S.N. Sinha, Dr. Dharmendra Singh, Dr. N.P. Pathak, Dr. Amalendu Patnaik, Dr.-Ing. Jagdish C. Mudiganti, Mr. Arun Kumar, Mr. Ashwini Arya, Ms. Ragini Jain, Mr. Abhay Gaekwad, Mr. B. B. Gupta, and Mr. Arjun Kumar for their support, suggestions and fruitful discussions throughout the work and during the writing. The first author also thanks the management of A. D. Patel Institute of Technology for their support during writing of this book. He also thanks his friends and colleagues at A. D. Patel Institute of Technology for their continual support during the work. Special thanks are due to Professor Dr. Manfred Thumm for forwarding the book and for his kind support. We are grateful to Dr. L.M. Joshi (CEERI, Pilani) for providing necessary data to carry out window calculations.

Thanks are due to the Department of Science and Technology (DST, New Delhi), Indian Institute of Technology Roorkee (IITR) and Central Electronics Engineering Research Institute (CEERI, Pilani) for their generous support and encouragement.

IIT–Roorkee   NCC
September 2011   MVK
                                                                                         AM

Special Acknowledgements

We sincerely thank the authorities of IIT–Roorkee for their kind permission to come up with this book. The first author also thank management of A. D. Patel Institute of Technology for their support in all aspects during the work. In addition, special thanks are due to all the authors of the original sources for the use of their work in the book.

We sincerely thank the following journals/publications/publishers of conference proceedings for their kind permission and for the use of their works and for reprint permission (publication details are given in the respective references and the corresponding citations are duly referred in the captions):

- IEEE, USA.

- Springer+Business Media B.V.

- Prentice-Hall India, New Delhi, India.

- De Gruyter, Germany.

- Frequenz, Germany.

- International Linear Accelerator Conference (LINAC).

- International Conference on Advanced Computing Technologies (ICACT-2008).

- NeuralWare, Carnegie PA, USA.

Contents

Introduction

1.1 Engineering Design Optimizations

Since the past few decades, engineering design and optimization problems have proved to be promising and important areas of research [1, 2]. Many researchers, engineers, and practitioners in academics and industry face difficulties in understanding the role of optimization in engineering design. The goal of optimization is not only to achieve a feasible solution, but also to meet design objectives. In most engineering design applications, the basic goal is to minimize the cost of products/production or maximize the efficiency of production. The goal of the overall design process is also to address the issues such as modeling the process, handling the constraints, forming the objective, and some times handling multiple objectives which may be conflicting in nature. With the advancements in high speed computing technologies, the optimization process has become a part of Computer-Aided Design (CAD) methodology. In a nutshell, *optimization is a very powerful tool but it must be applied judiciously in order to achieve efficient solution in feasible time.* The contents in the book deals with soft computing methods for the design applications in microwave and millimeter-wave domain.

1.2 Soft Computing (SC) Methods

Soft computing is defined as a collection of computational techniques in computer science, artificial intelligence and related fields, which attempt to study, model, and analyze very complex phenomena for which more conventional (hard computing) methods have not obtained low cost, analytic, and complete solutions [3, 4]. The term soft computing was first introduced by Prof. L. A. Zadeh, University of California in 1992 [5]. The aim of soft computing is to tolerate imprecision, uncertainty, and approximation in order to achieve robust and low cost solution in a small time frame. Much of the soft computing techniques have been inspired from biological phenomena and the social

N. Chauhan, M. Kartikeyan, and A. Mittal: Soft Computing Methods, SCI 392, pp. 1–7.
springerlink.com

behavior of biological populations. Since so long time in the literature, the term of soft computing was confined to mainly three techniques such as Artificial Neural Network (ANN), Genetic Algorithm (GA), and Fuzzy Logic (FL). However, the major elements of soft computing include (but yet not confined to!) neural network, evolutionary computation, fuzzy logic, machine learning, and probabilistic reasoning [3]. The recently developed methods based on swarm intelligence, and foraging behavior of natural and biological populations such as birds, fishes, ants, and bacteria are also considered to be part of the growing field of soft computing.

1.3   Microwave and Millimeter-Wave Design

Major practical problems in the field of microwave and millimeter-wave domain are design problems. These problems include design of microwave components such as antenna, filter, resonator, coupler, transmission lines, bends, and so on. Each component may be designed for different applications and with different techniques involved with it. There design problems are of various types and each is having different challenge with it. The growing commercial market of wireless communication devices over the past decade has also led to the explosion of interest and opportunities for design and development of RF and microwave components. The microwave industry emphasizes on the development of these components and systems in the shortest possible time and at low development cost [6]. This places the demand on various CAD tools for the development of microwave components. There is also a class of critical design problems (mainly from the millimeter-wave domain) where the reliability and accuracy in the design are the prime requirements.

The steps of a conventional process followed for the design of microwave and millimeter-wave components consists of steps such as problem identification, specification generation, concept development, electromagnetic analysis, evaluation, initial design, final design, fabrication, and testing [6]. At present, the design of most microwave components is carried out using commercially available Electromagnetic (EM) simulation tools such as IE3D, HFSS, Microwave Studio, etc. Many new EM simulation tools are being developed to automate the design process. EM simulation techniques help to produce EM analysis for microwave components to be designed. In the design process, the role of simulators is to obtain responses such as S-parameters, standing wave ratios, gain, current distributions, power transmission and reflections, etc. In the conventional design methodology, once an initial design is obtained, the EM analysis and evaluation is performed iteratively until the desired specifications are met. In this process, one has to change the design parameters by modifying its geometrical structure and apply expert domain knowledge to make the design feasible, and to move towards desired objective. This process is repeated till a tolerable/acceptable solution is achieved. But it does not guarantee for the optimum solution. This approach may, sometimes, degrade the performance of the component after its fabrication in the desired subsystem. Moreover, these methods of designing and optimizing them by hand are laborious, time intensive, and require designers to have significant knowledge about electromagnetics, microwave engineering, and other specialized subjects concerning the design. Eventually, it is necessary to use various optimization algorithms to reach the optimum parameters. Some of the present EM simulators use conventional (mathematical) optimization methods [2] like golden search method, steepest descent method, conjugate gradient method, quasi-Newton method, and other random search methods for optimization of the design parameters. The difficulties with these local search methods are that they require a proper initial guess; otherwise the chances of getting local optimum solutions are very high. Moreover, they can only handle a very few number of design parameters. In addition, handling of many design constraints at the same time is difficult. This conventional way of microwave design has also achieved a certain level of maturity in recent years. To make the efficient use of EM simulation techniques in the development of microwave components is still a topic of research. New techniques are required in order to search from a large design space and reach an optimum solution.

Soft Computing (SC) methods offer unique advantages for the design problems of different disciplines including microwave and millimeter-wave domain problems. These advantages are listed as below.

- The SC methods can be easily interfaced with many of the EM simulators due to which the laborious task of optimizing design parameters in manual mode can be converted to computer simulations. Hence, faster results can be obtained. Moreover, the use of global optimization methods can reduce the chances of obtaining local optima.
- Handling many design constraints simultaneously becomes easy with SC methods.
- The use of SC methods does not require extensive mathematical formulation of the problem. Thus the requirement on the necessity of exclusive domain specific knowledge can be reduced. This can permit designers from other disciplines to work in the area of electromagnetic and microwave design.
- Since most of the SC methods come under "GNU general public license" (open source), they provide low cost solutions to the designers. Moreover, effective hybridization of SC methods may also reduce the dependency on costly EM simulators up to some extent.
- SC methods are adaptive and scalable. Though in this book, we discuss SC methods for microwave and millimeter-wave design problems, they are applicable to the design applications of many other engineering disciplines.

Soft computing methods such as GA and ANN have been widely exploited by EM researchers for microwave design since last decade [6–15, 64, 16–20]. Though these techniques are effective, they suffer from certain drawbacks. Evolutionary algorithms such as GA is effective in finding optimum solution from a multivariate feature space, but it is iterative and require large amount of computation to reach an optimum solution. ANN has been widely used for fast microwave modeling but it has also suffered with generalization problems. This has created the need to develop novel modifications and hybridizations of SC methods in order to solve microwave design problems effectively. Moreover, during the recent past, several new soft computing methods such as Bacterial Foraging Optimization (BFO) [21], Ant Colony Optimization (ACO), Support Vector Machines (SVM), Artificial Immune System (AIS) [22], etc., have emerged along with their applications in engineering design and optimization problems. Many of these recent and other SC methods have not been yet investigated or very few works have been reported on these methods for microwave design tasks. This also motivates investigation of these methods for the design tasks in microwave domain.

1.4   Challenges, Objectives and Scope

1.4.1   Design Challenges

Although there are various EM simulation tools and soft computing methods, the following design issues and requirements make the field of microwave and millimeter-wave design a true challenging aspect:

Sensitivity: The key challenge in the design of microwave components is to deal with sensitivity. Due to very small wavelengths involved in the microwave and millimeter-wave regions, any small variation in the physical aspects of the components may result in large variation in the output response. This makes the design surface extremely non-smooth. Thus, modeling and finding an optimum solution make the design task challenging. It necessitates fine tuning of design parameters in the feature space. Moreover, increasing number of design parameters make the feature space too large to explore. Hence, in order to achieve high precision requirements, unconventional methods have to be employed in the design process.

Size reduction and low weight: Due to emerging technologies and miniaturization of the components, optimization is required with high precision.

Time-to-market: The recent trend in industry suggest the design and manufacturing of components in small time-frame to satisfy the need of growing commercial market. Therefore, the design and development should be as fast as possible.

Low cost: The EM simulators available in the market for designing components are very costly. It is not possible for an industry or institute to purchase all the tools. This also demands the use of efficient alternative solutions for the design of components. The process adopted should be such that it reduces the cost of production.

Reliability: Microwave/millimeter-wave components are extensively used in critical areas such as communications, radar, defense, and some specific industrial–scientific–medical (ISM) applications where the reliability of the component is the prime requirement. Malfunctioning of one component may cause the failure of the entire system and spell catastrophic damage. It is required to develop and investigate soft computing methods that provide accurate and reliable designs.

1.4.2 Overall Objectives

The objectives covered in the different chapters of this book can be stated as follows [23]:

- To demonstrate use of some soft computing based methods that lead to faster design of microwave components.
- To demonstrate soft computing based methods that lead to accurate design for certain critical applications such as design of components for high power millimeter-wave sources.
- Effective hybridization of soft computing methods with EM simulators and also with each other to solve design problems in microwave field.
- To help readers and practitioners learn effectively from empirical data and develop empirical models of microwave components via support vector machine framework.
- To present a state-of-the-art review on the use of soft computing methods for microwave design tasks and to present the investigation of latest soft computing methods for microwave design.

In order to meet the above mentioned objectives and demonstrate the potential strength of the presented soft computing methods, a pack of diversified problems with varied challenges have been demonstrated throughout the book.

These design problems are as follows:

- Design of a coupled microstrip-line band pass filter
- Efficient modeling of a simple one-port microstrip via
- Design of a circularly polarized microstrip antenna and a simple aperture coupled microstrip antenna
- Design of nonlinear tapers for specific high power gyrotrons
- Design of disc type RF-windows for high power millimeter-wave sources such as gyrotrons and klystrons

The examples of microstrip antennas, and microstrip filter are mainly considered to demonstrate different optimization approaches for faster design. Again, the efficiency of SVM based modeling is demonstrated by a microstrip via, and two microstrip antennas. Whereas, the critical components - nonlinear tapers and RF-windows - are chosen to show the ability of different soft computing methods to obtain precise and accurate designs where the tolerance is very less. It should be noted that the applications considered here are case-study design problems, but the methods presented can well be applied to many other microwave design problems.

The overall objective of the presented concepts and experiments was to develop and investigate soft computing based methods for tackling challenges in the design problems of microwave domain. Our aim in the development and investigation of these methods was to obtain the designs in small time frame while maintaining or improving the efficiency of the design process for a wide range of applications. One of the goal is also to learn effectively from empirical data, and to provide a framework for efficient modeling of microwave components.

Thus the methods presented here emphasizes on dealing with challenges in solving diversified design problems of microwave field using soft computing based methods. The contents presented in this book addresses the engineers, practitioners, and researchers in the field of soft computing, modeling and optimization in engineering, and RF/microwave design.

1.4.3 Scope

A typical microwave design process consists of problem identification, specification generation, concept development, EM analysis, evaluation, initial design, final design, fabrication, and testing. Out of these steps, the presented work in the book is focused on converting initial design to final design which includes steps such as modeling, computer-aided analysis, and optimization. Thus, the assumption is made that the initial designs and some information about the design parameters (such as specifications, ranges, etc.) are available. In the presented work, this information was obtained from previous published works and expert knowledge. Moreover, as the presented work emphasizes on soft computing methods for microwave and millimeter-wave design problems, the fabrication of the components is not considered. It may be noted that the term 'experiment' or 'experimental' is used to refer 'simulation' and not the physical experiment. It should also be noted that the soft computing methods presented here do not aim to replace the use of EM simulators from microwave design, but they try to overcome the shortcomings present in the conventional design methodology.

1.5 Outline of the Book

The organization of the later chapters is as follows.

Chapter two of the book provides concepts, algorithms and other mathematical foundations for the different soft computing methods in general. These concepts are used in later chapters to solve variety of microwave and millimeter-wave design problems.

Chapter three presents a state-of-the-art review on the present use of soft computing methods for design applications in microwave domain.

In chapter four, a modified particle swarm optimization algorithm with a novel concept of multiple sub-swarms is presented and its applicability is demonstrated by the designs of a specific microwave filter.

Chapter five of the book presents efficient modeling of microwave components using SVM and also presents a hybrid approach - support vector driven evolutionary algorithm - for designing microwave components such as a microstrip via, and two microstrip antennas.

Chapter six presents the design and optimization of a nonlinear taper for a specific high power gyrotron using two swarm intelligence based algorithms namely a modified bacterial foraging optimization and a standard particle swarm optimization.

Chapter seven presents the design of disc-type RF-windows using multi-objective particle swarm optimization methodology.

Finally in chapter eight, the concepts presented in the book are summarized and scope of the future work is outlined.

Soft Computing Methods

2.1 Overview

In this chapter, a brief introduction to different soft computing methods, namely, genetic algorithms, particle swarm optimization, bacterial foraging optimization, neural networks and support vector machine has been presented. The overall presentation is separated into sections like evolutionary algorithms, swarm intelligence based algorithms, and methods of learning. Among these methods, genetic algorithms and neural networks have been well explored by microwave researchers, however rest of the techniques are still not much investigated. The rest of the book chapters focuses much on the investigation of these relatively new techniques and their modifications to microwave and millimeter-wave design problems.

2.2 Evolutionary Algorithms

Evolutionary algorithms includes a set of problem-solving techniques based on the principles of biological evolution. The main components of evolutionary computation includes: genetic algorithms, genetic programming, evolutionary strategies, evolutionary programming. The detailed discussion of all four evolutionary computation techniques is beyond the scope of the book. In this section, the basic principle of genetic algorithms is given. It is one of the most popular nonconventional population based optimization algorithm.

2.2.1 Genetic Algorithms

Genetic Algorithms (GA) [24–26] are evolutionary search and optimization algorithms based on the mechanics of natural selection and natural genetics. GA is different from most traditional optimization methods. The major difference between GA and other traditional search methods is that it starts with a set of solutions in contrast to the single solution approach used by

N. Chauhan, M. Kartikeyan, and A. Mittal: Soft Computing Methods, SCI 392, pp. 9–23.
springerlink.com   © Springer-Verlag Berlin Heidelberg 2012

traditional optimization methods. Each solution in GA is known as *chromosome* and the set of solutions (chromosomes) is known as *population*. In the evolution process of GA, solutions from one population are taken and used to form a new population with a hope that the new population will be better than the older one. Defining objective function (fitness function), representing chromosomes and applying genetic operators are three most important aspects of GA. The power of GA in effectively searching a multidimensional search space lies in its three basic operators: reproduction, crossover, and mutation.

The four main important components concerning genetic algorithm includes, fitness function, number and type of variables, problem constraints, and number of objective functions to be optimized.

- *Fitness function:* The fitness function (or objective function) is a problem dependent function $f(x)$ to be optimized. The GA normally works for minimization problems. However, the maximization problem can be converted to minimization problem with the help of duality principle, as follows:
  Minimize $- f(x)$
  or Minimize $1/f(x)$, for $f(x) \neq 0$
  or Minimize $1/(1 + f(x))$, for $f(x) \geq 0$
  or Minimize $1/(1 + \{f(x)\}^2)$, and so on.
- *Design parameters:* Based on the number of variables one or more, the problem can be categorized to be univariate or multivariate design problem. In the basic form the variables in GA are used in binary form, and hence it is known as binary-coded GA. More general version of GA that allows to work with real variables is real-coded GA.
- *Constraints:* The optimization problem can also be classified based on the number and type of constraints. The problem without any explicit constraints is known as unconstrained problem. However, this may include boundary constraints of design variables. The boundary constraint is a range of design parameter from which the solution is desired to be found. The constraints for a constrained optimization problem can be linear or nonlinear, equality or inequality. Depending on the type of constraints different constrained handling methods can be involved.
- *Number of objective functions:* For most optimization problems, there is single objective function to be optimized. However, a problem may have multiple objectives to be optimized simultaneously. This kind of problem is known as multi-objective optimization problem. In this case, multiple solutions can be found based on the trade-off between multiple objectives. Further discussion on multiobjective optimizations is given in chapter 7 of the book.

Algorithm

The outline of the standard genetic algorithm is as follows:

ALGORITHM: *Genetic Algorithm*

1. *Initialization:* Generate random population of n solutions (encoded by chromosomes) to a given problem.
2. *Fitness evaluation:* Evaluate the fitness $f(x)$ of each chromosome x in the population.
3. *New population generation:* Generate a new population (offsprings) from old population by applying following steps,

    a. *Reproduction:* Chromosomes are selected from parents, according to some selection strategy, to reproduce and generate offsprings. This selection is carried out in a probabilistic manner according to their fitness value. This follows Darwin's evolution theory of survival of fittest (the better fitness, the bigger chance to be selected). Reproduction makes clones of good chromosomes, but does not create new ones.
    b. *Crossover:* Crossover proceeds in three steps: selection of a pair of chromosomes for mating based on some probability, selection of cross-site, and swapping the genetic material between two chromosomes based on their cross-site. The aim of crossover is to search parameter space.
    c. *Mutation:* The mutation (changing bit from 0 to 1 or vice-versa) of the chromosomes is performed based on mutation probability. The mutation operator preserves the diversity among the population, prevent premature convergence, and restores lost information to the population. The probability of mutation is generally too low compared to probability of crossover.
    d. *Fitness evaluation:* Evaluate the fitness for newly generated population.

4. *Optimization loop.* Repeat steps 3 and 4 until some predefined error criteria or maximum number of iterations are attained.

2.3 Swarm Intelligence Based Methods

2.3.1 Particle Swarm Optimization

Standard PSO

The standard PSO is a population based algorithm. Each potential solution in the population is known as *particle*, which is represented by position and velocity vectors. The particles fly through the multidimensional search space in order to get the best solution. These particles adjust their velocity according to their own flying experience and according to the experience of their companions. Let, the position and velocity of each particle in the *swarm* (a population of particles) are represented as $X_i = (x_{i1}, x_{i2},..., x_{iD})$, and

$V_i = (v_{i1}, v_{i2},..., v_{iD})$ (where D is number of decision parameters of an optimization problem) respectively. The best previous position of each particle is represented as $P_i = (p_{i1}, p_{i2},..., p_{iD})$. The global best position of all particles is represented by $P_g = (p_{g1}, p_{g2},..., p_{gD})$. The velocity and position of each particle are updated using relations:

$$v_{id} = v_{id} + c_1 r_1 (p_{id} - x_{id}) + c_2 r_2 (p_{gd} - x_{id}) \qquad (2.1)$$

and

$$x_{id} = x_{id} + v_{id} \qquad (2.2)$$

where c_1 (cognitive constant) and c_2 (social constant) are two positive constants, r_1 and r_2 are two random numbers uniformly generated between $[0, 1]$. The values of c_1 and c_2 are considered to be equal in most PSO literatures to balance movement of particle in both cognitive and social components. The velocities of particles are clamped by maximum velocity vector V_{max} on each dimension. The pseudo-code for standard PSO is shown in Fig. 2.1. The global best version (GBEST) of standard PSO [27, 28] is considered for the experiments in this book.

```
Initialize positions and velocities of all particles in the swarm randomly
Repeat
    For each particle in the swarm
        Calculate the fitness value f(Xᵢ)
        If f(Xᵢ) < f(Pᵢ) then Pᵢ = Xᵢ
    End for
    Update Pg, if the best particle in the current swarm has lower f(X) than f(Pg)
    For each particle in the swarm
        r₁=random(), r₂=random()
        Calculate particle velocity according to equation (2.1)
        Restrict the velocity of particles by [−Vmax, Vmax]
        Update particle's position according to equation (2.2)
    End for
Until maximum iteration or minimum error criteria is attained
```

Fig. 2.1 Pseudocode of standard PSO algorithm [27, 28]

Variations of PSO

There are several variations to the standard PSO algorithm, a comprehensive summary of which is given in [29]. In this section, we discuss two basic variations of PSO. The first is PSO with Inertia Weight Method (IWM) which is proposed by Shi and Eberhart [30, 31], while the second is PSO with Constriction Factor Method (CFM) which is proposed by Clerc [32].

Shi and Eberhart [30] introduced Inertia Weight (IW) parameter into original particle swarm optimizer. The purpose was to balance exploration and exploitation abilities of PSO. Basically, IW controls the momentum of the particle by weighing the combination of previous velocity. The velocity update Eq. (2.1) is modified to,

$$v_{id} = wv_{id} + c_1r_1(p_{id} - x_{id}) + c_2r_2(p_{gd} - x_{id}), \qquad (2.3)$$

where w represents inertia weight which is used to balance local and global search abilities. Different ways of selecting value of inertia weight are given in literature [29], which include linearly decreasing IW, nonlinearly decreasing IW, linearly increasing IW, and fuzzy adaptive IW. In this book, we consider a common choice of considering linearly decreasing inertia weight from an initial value of 0.9 to final value 0.4.

In another variation, Clerc [32] proposed a different approach to balance the trade-off between exploration and exploitation. He introduced a constant which is referred as *constriction coefficient* to constrict the velocity of particles. The velocity update Eq. (2.1) is modified to,

$$v_{id} = \chi[v_{id} + c_1r_1(p_{id} - x_{id}) + c_2r_2(p_{gd} - x_{id})], \qquad (2.4)$$

with constriction factor χ defined as

$$\chi = \frac{2\kappa}{|2 - \phi - \sqrt{\phi(\phi - 4)}|}, \phi = c_1 + c_2, \phi > 4, \qquad (2.5)$$

where $\kappa \in [0,1]$ is a positive constant. Here parameter κ controls the exploration and exploitation abilities of the swarm. A comparison of IWM and CFM is given in [33]. It was concluded that CFM has relatively better convergence abilities than IWM.

In chapter 4, we present a modified PSO considering multiple subswarms and apply it to both the above variations of PSO. The proposed algorithm (after adapting above modifications) is tested with five benchmark functions commonly used in PSO literatures. The results of modified PSO with above two variations (IWM and CFM) are compared with standard PSO with same variations respectively.

2.3.2 Bacterial Foraging Optimization

Bacterial foraging algorithm is inspired by the pattern exhibited by foraging behavior of bacteria, more specifically *Escherichia coli* (E. coli) bacteria, which resides in our intestines [34]. The algorithm proceeds by either selecting or eliminating bacteria based on their good or poor foraging strategies. Further, the algorithm may also refine poor foraging strategies into better ones.

Standard Bacterial Foraging Optimization

The foraging process in BFO algorithm involves four processes such as chemotaxis, swarming, reproduction, and elimination-dispersal. These four processes are described as follows:

Chemotaxis: The movement of E. coli bacteria towards the nutrient-rich areas is simulated by an activity called "chemotaxis". This process involves two sub-activities: tumbling and swimming. In tumbling, bacteria do not move noticeably, but positions themselves in some random directions in which later on swimming can occur. Swimming is performed in the specified direction with fixed swim length. The process of chemotaxis can be represented as,

$$\theta^i(j+1,k,l) = \theta^i(j,k,l) + C(i)\phi(i) \tag{2.6}$$

where $\theta^i(j,k,l)$ indicates the position of the i-th bacterium at the j-th chemotactic step, in the k-th reproductive loop, and at l-th elimination-dispersal event. $\phi(i)$ is a random unit vector, and $C(i)$ is the length of a unit walk in the direction specified by $\phi(i)$.

Swarming: It is a group behavior or cell-to-cell signaling exhibited by bacteria while moving towards rich-nutrient areas. In this, the healthy bacteria attract other bacteria towards them. Bacteria move into a group forming a pattern. Mathematically, swarming can be represented as,

$$
\begin{aligned}
J_{CC}(\theta, P(j,k,l)) &= \sum_{i=1}^{S} J_{CC}^i(\theta, \theta^i(j,k,l)) \\
&= \sum_{i=1}^{S}[-d_{attract}\exp(-w_{attract}\sum_{m=1}^{p}(\theta_m - \theta_m^i)^2)] \\
&+ \sum_{i=1}^{S}[h_{repellent}\exp(-w_{repellent}\sum_{m=1}^{p}(\theta_m - \theta_m^i)^2)]
\end{aligned}
\tag{2.7}
$$

where $J_{CC}(\theta, P(j,k,l))$ is the cost function value added to the actual cost function which makes it time varying. Here, S indicates number of bacteria, p indicates number of design parameters, and $\theta = [\theta_1, \theta_2, ..., \theta_p]^T$ is a point in the p-th dimensional search space, while $d_{attract}$, $w_{attract}$, $h_{repellent}$, and $w_{repellent}$ are constants and should be chosen carefully.

Reproduction: In this process, healthy bacteria reproduce and split into two, while unhealthy bacteria die. For convenience, the total population is maintained constant by killing half the population with highest fitness values, and remaining half reproduces and is placed at same locations.

Elimination-dispersion: In this event, some bacteria are dispersed to random locations with some probability due to factors such as consumption of food, and other environmental effects. The bacteria with probability p_{ed} are dispersed to random locations in the optimization domain.

The flowchart of the standard Bacterial Foraging Optimization (BFO) algorithm is shown in Fig. 2.2, while the detailed algorithm [34] is stated as below.

ALGORITHM: *The Bacterial Foraging Optimization Algorithm*

Initialization:

1. Initialize parameters Dim, S, N_c, N_s, N_{re}, P_{ed}, N_{ed}, $C(i)$ with $(i = 1, 2, ..., S)$, θ^i where,

   Dim: Dimension of the search space,

   S: Number of bacteria in the population,

   N_c: Number of chemotactic steps,

   N_{re}: Number of reproduction steps,

   N_s: Length of swimming,

   N_{ed}: Number of elimination-dispersal events,

   P_{ed}: Probability of elimination-dispersal events,

   $C(i)$: Size of the step taken in the random direction specified by the tumble,

   $\theta^i(j, k, l)$: Position vector of the i-th bacterium, in j-th chemotactic step, k-th reproduction step and l-th elimination-dispersal event.

Iterative loops:

2. Elimination-dispersal loop: $l=l+1$
3. Reproduction loop: $k=k+1$
4. Chemotaxis loop: $j=j+1$

   a. For $i = 1, 2, ..., S$, take a chemotactic step for bacterium i as follows:
   b. Compute fitness function $J(i, j, k, l)$, and then let,
      $J(i, j, k, l) = J(i, j, k, l) + J_{CC}(\theta^i(j, k, l), P(j, k, l))$.
   c. Let $J_{last} = J(i, j, k, l)$ to save this value since we may find a better cost via a run.
   d. *Tumble:* Generate a random unit vector $\phi(i)$ with each element $m(i)$, $m=1, 2,..., Dim$, a random number on $[-1, 1]$.
   e. *Move:* Following the Eq. (2.6).
   f. Compute $J(i, j+1, k, l)$ as, $J(i, j+1, k, l) = J(i, j+1, k, l) + J_{CC}(\phi^i(j+1, k, l), P(j+1, k, l))$.
   g. *Swim:* Consider only the i-th bacterium is swimming while the others are not moving, then,
      i. Let $m = 0$ (counter for swim length)
      ii. While $m < N_s$ (if have not climbed down too long)
         • Let $m=m+1$
         • If $J(i, j+1, k, l) < J_{last}$ (if doing better) then,
            $J_{last} = J(i, j+1, k, l)$ and $\theta^i(j+1, k, l) = \theta^i(j+1, k, l) + C(i)\phi(i)$ and use this $\phi^i(j+1, k, l)$ to compute the new $J(i, j+1, k, l)$ as in (f).
         • Else, if $m = Ns$ then end while loop.

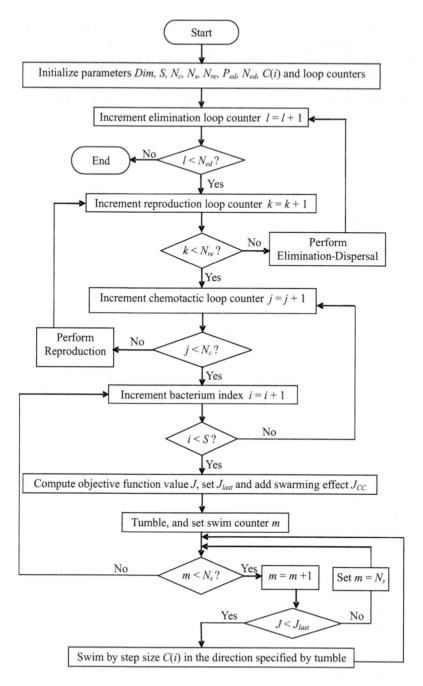

Fig. 2.2 The flowchart of standard BFO algorithm [34] (© 2002 IEEE)

h. If $i \neq S$ then $i = i + 1$ and go to (b) to process $(i+1)$ bacterium.

5. If $j < N_c$ go to step 4 (i.e., continue chemotaxis as the life of the bacteria is not over).

6. *Reproduction:* Sort bacteria in ascending of their fitness value (J).
   Now, let $S_r = S/2$. The S_r bacteria with highest cost function (or fitness) values (J) die and the other half of bacteria population with the best values split (and the copies that are made are placed at the same location as their parent).

7. If $k < Nre$, go to step 3. We have not reached the specified number of reproduction steps. So we start the next generation of the chemotaxis loop.

8. *Elimination-dispersal:* For $i = 1, 2, ..., S$, eliminate and disperse each bacterium with probability P_{ed}. (If any bacterium is eliminated, then disperse other bacterium to random location in optimization domain in order to keep the number of bacteria in population constant.) If $l < N_{ed}$ then, go to step 2; otherwise end.

2.4 Soft Computing Methods for Modeling and Approximation

2.4.1 Artificial Neural Networks

Artificial Neural Network (ANN) [35, 36, 26] is a simplified model of the biological neuron system. It can be considered as a parallel distributed processing system of highly interconnected computing elements called neurons. It acquires knowledge by learning from empirical data and makes it available for use for solving a given problem. Acquiring the knowledge through past data is known as training and the ability to solve a problem using the knowledge acquired is known as inference.

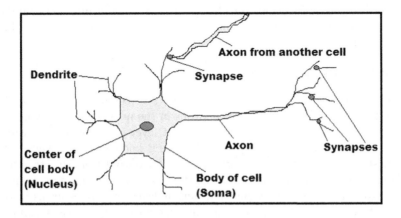

Fig. 2.3 A biological neuron [37]

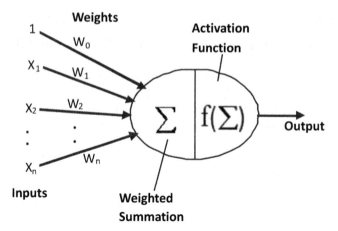

Fig. 2.4 Simple model of an artificial neuron [26] (Source: S. Rajasekaran and G.V. Vijayalakshmi Pai, Neural Networks, Fuzzy Logic and Genetic Algorithms - Synthesis and Applications 2011 - Reproduced with permission of PHI Learning Private Limited, New Delhi)

The model of a biological neuron is shown in Fig. 2.3. It is composed of cell body (soma), axon, and hair-like dendrites. The axon splits into branches terminating in very small bulbs that almost touch with the dendrites of other cell. The small gap between terminating bulb and a dendrite is called a synapse. It is the synapse through which information is propagated from one cell to another.

The general model of artificial neuron representing the terminologies of biological neuron is shown in Fig. 2.4. Here, the synapse is represented by connection and synaptic efficiencies between biological neurons is represented through weights. The neuron cell plays roll of summing junction followed by activation function. Various activation functions used in neural network structures are linear function, threshold function, signum function, sigmoid function, hyperbolic tangent function, etc. These transfer functions are illustrated in Fig. 2.5.

Neural Network Architecture

Neural network structure is represented using a directed graph $G = (V, E)$ of V vertices and E edges. Different neurons in the neural network structure are organized in different layers. The left most layer contains neurons which does the task of receiving the input, and hence is known as input layer. The right most layer contains neurons which generates output from the structure, and hence is known as output layer. The structure may include zero or more hidden layers lying between input and output layer. All the neurons in the previous layer are connected to all the neurons in the next layer forming a feed forward network. This network does not contain any connection from

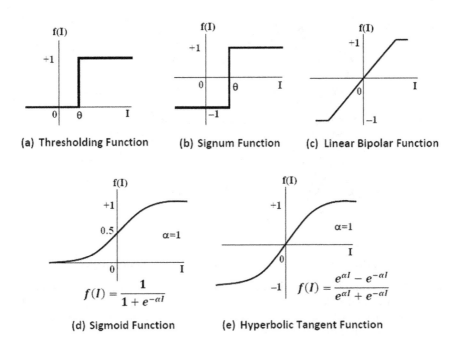

Fig. 2.5 Various activation functions [26]

one layer to the same or previous layer. The other type of network, known as recurrent network, contains atleast one feedback connection.

One of the most popular neural network architectures used in wide range of applications is feedforward neural network. The architecture of a multilayer feedforward neural network is shown in Fig. 2.6. It allows the signals to move only in forward direction from the input to the output layer through hidden layers without making feedback loops.

Learning Methodology

The neural network should be configured in such a way that the set of inputs for a given application generates desired set of outputs. The neural network is trained by giving appropriate input-output pairs. During this training weights are adjusted according to some learning mechanism. The neural network learning can be categorized in three types: supervised learning, unsupervised learning and reinforcement learning.

Supervised learning: In this learning process, a training pattern from the training set is applied randomly to the network, error is calculated based on the desired response and the network synaptic weights are adjusted in order to minimize the error. Here, a teacher is assumed to be present behind the learning process.

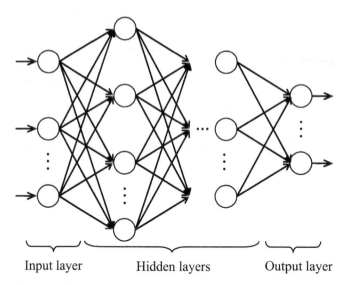

Input layer   Hidden layers   Output layer

Fig. 2.6 An example of a multilayer feedforward neural network [26] (Source: S. Rajasekaran and G.V. Vijayalakshmi Pai, Neural Networks, Fuzzy Logic and Genetic Algorithms - Synthesis and Applications 2011 - Reproduced with permission of PHI Learning Private Limited, New Delhi)

Unsupervised learning: In this learning process, no output is presented to the network as if no teacher is available to present the desired response. Here, only some data and an objective function which is to be optimized are available during the learning.

Reinforcement learning: In this learning process, a teacher is present but do not specify the exact output, rather he gives reward for the correct answer and penalty for the wrong answer.

Backpropagation Learning Algorithm

Backpropagation learning is one of the most popular supervised learning methods for training multilayer feedforward neural network architecture. The steps followed by backpropagation learning algorithm are stated as follows:

ALGORITHM: *Backpropagation Learning Algorithm*

1. *Determine the structure of the neural network and other parameters:* This includes determining number of nodes in the input layer, number of nodes in the output layer, number of hidden layers, number of nodes in each hidden layer. There are also some other important parameters such as learning rate, momentum, number of training samples, etc., that are required to be determined carefully.
2. *Initialization of weights:* Initialize the network synaptic weights to small random values.
3. Repeat

   a. For each training pattern
     i. *Forward pass:* The input patterns are applied to the input layer nodes which simply forward them to the input of the first hidden layer. The computation of input and output is performed at each hidden and output layer nodes.
     ii. *Calculate the error in the output layer:* Based on the output available at output layer node and the values of desired output, error is calculated at each output layer nodes.
     iii. *Error backpropagation and modification of weights:* In this step, the error at the output node is propagated back in the network and the synaptic weights of the network are updated based on the errors available at each of its posterior nodes.
   b. end

4. Until the desired termination criteria is attained

2.4.2 Support Vector Machine

Conceptual Background

Support vector machine (SVM) is a machine learning technique developed by Vapnik [38, 39]. It is based on the principle of statistical learning theory. SVM was initially developed for pattern recognition task but later on its application was extended to regression problems. Support Vector Regression (SVR) is found to give robust and effective model of the process under consideration [40]. The models developed using SVR are simple and their evaluation is fast. No prior knowledge about input/output mapping is required for the model development. An increasing number of engineers and researchers from diverse fields have begun to take a serious interest in this emerging technology. The hypothesis generated using SVM involves both structural risk minimization (SRM) and empirical risk minimization (ERM). This makes SVM much more powerful in generalizing than traditional ANN which only minimizes empirical risk. The key ideas of SVM are: nonlinear mapping from input space to high-dimensional feature space using a kernel function and finding an optimum hyperplane that maximizes generalization ability [38, 40].

Support Vector Regression

Learning systems for SVR estimation is described in [40–42]. Given a set of input-output training data $(x_i, y_i) \in R^n \times R$, $i=1$ to l, we need to estimate a function $f : R^n \to R$ that will correctly predict unseen examples generated from the same underlying probability distribution as the training data. The hypothesis of SVM maps the original input space into a high dimensional space via a kernel. This higher dimensional space is called *feature space*, in which an optimal hyperplane is determined to maximize the generalization ability [43]. The generic support vector regression estimation function takes the form:

$$f(x) = (w\Phi(x) + b) \tag{2.8}$$

where, $w \in R^n, b \in R$ and Φ denotes a nonlinear transformation from R^n to a high dimensional feature space. Our goal is to find the value of w and b that minimize the regression risk defined as:

$$R(f) = \frac{1}{2}\|w\|^2 + C\sum_{i=0}^{l} \Gamma(f(x_i) - y_i) \tag{2.9}$$

where Γ is the cost function, C is a constant, and vector w can be written in terms of data points as:

$$w = \sum_{i=1}^{l} (\alpha_i - \alpha_i^*)\Phi(x_i). \tag{2.10}$$

The ε-insensitive loss function [43] is the most widely used cost function. This function is of the form:

$$\Gamma(f(x) - y) = \begin{cases} 0, & \text{for} \quad |f(x) - y| \le \varepsilon \\ |f(x) - y| - \varepsilon, & otherwise. \end{cases} \tag{2.11}$$

The regression risk in Eq. (2.9) and the ε-insensitive loss function in Eq. (2.11) can be minimized by maximizing the following dual optimization problem with respect to α and α^*,

$$W(\alpha, \alpha^*) = \sum_{i=1}^{l} y_i(\alpha_i - \alpha_i^*) - \varepsilon \sum_{i=1}^{l}(\alpha_i + \alpha_i^*) - \frac{1}{2}\sum_{i,j=1}^{l}(\alpha_i - \alpha_i^*)(\alpha_j - \alpha_j^*)K(x_i, x_j)$$
$$\tag{2.12}$$

with constraints

$$\sum_{i=1}^{l}(\alpha_i - \alpha_i^*) = 0, \quad where \quad \alpha_i, \alpha_i^* \in [0, C], \quad i = 1, .., l. \tag{2.13}$$

Here, α_i and α_i^* are Lagrange multipliers which represent solutions to the above quadratic problem. Only the nonzero values of the Lagrange multipliers in Eq. (2.12) are useful in forecasting the regression curve and are known as *support vectors*. The constant C introduced in Eq. (2.9) determines penalties to estimation errors. A large C assigns higher penalties to errors so that SVM is trained to minimize error with lower generalization, while a small C assigns fewer penalties to errors. This allows the maximization of margin with errors, thus higher generalization ability.

The approximation for nonlinear data set is accomplished with the use of kernel functions. According to Christianini and Shawe-Taylor [41], one of the most important design choices for SVM is the kernel-parameter, which implicitly defines the structure of the high dimensional feature space where a maximal margin hyperplane is found. Too rich a feature space would cause the system to overfit the data, and conversely the system might not be capable of approximating the data if the kernels are too poor.

3

Review of Soft Computing Methods for Microwave and Millimeter-Wave Design

3.1 Overview

In this chapter, we present a state-of-the-art review on the present use of soft computing methods for the design applications in microwave and millimeter-wave domain [44]. Since long time, the literature on soft computing was confined to the methods such as genetic algorithms, artificial neural network, fuzzy logic, and their variations and hybridizations. During last decade, few other swarm intelligence based algorithms such as particle swarm optimization, ant colony optimization, and bacterial foraging optimization have emerged. Microwave researchers also observe these techniques and try to adopt them for various microwave design applications. In this chapter, a review of microwave design using five soft computing methods namely Genetic Algorithm (GA), Particle Swarm Optimization (PSO), Bacterial Foraging Optimization (BFO), Artificial Neural Network (ANN), and Support Vector Machine (SVM) has been presented. Out of these methods, ANN and GA have been widely exploited by microwave researchers. Though efforts have been made to review related works of all five methods used for microwave design applications, emphasis is given on recent methods, namely, PSO, SVM and BFO. For BFO and SVM, no much has been reported in literature for microwave design.

For convenience, we have divided the discussion of the review into four parts: microwave design using genetic algorithms, microwave design using SI based algorithms such as PSO and BFO, microwave modeling using soft computing methods such as ANN and SVM, and hybridization of soft computing methods used for microwave design. The list of references covered in this review is by no means exhaustive, but it is fairly representative of present usage of these techniques.

The organization of the chapter is as follows. Section 3.2 presents microwave design using genetic algorithms. Section 3.3 presents microwave and millimeter-wave design using swarm intelligence based algorithms namely PSO and BFO. Section 3.4 discusses microwave modeling using two

N. Chauhan, M. Kartikeyan, and A. Mittal: Soft Computing Methods, SCI 392, pp. 25–33.
springerlink.com © Springer-Verlag Berlin Heidelberg 2012

techniques, namely, ANN and SVM. Section 3.5, presents a review on the usage of hybrid soft computing techniques for microwave design.

3.2 Microwave Design Using Genetic Algorithms

3.2.1 Benefits of Microwave Design Using Genetic Algorithms

Currently used methods of microwave design suggest the use of full-wave analysis of EM simulation tools. Many new EM simulation tools are being developed by industry to automate the design process. Some of them are embedding local search methods for optimizing the design parameters. Evolutionary Algorithms (EAs) such as genetic algorithms are reliable alternatives to these methods for getting optimum designs.

There are many advantages of using EAs for engineering design including microwave field, some of them are as follows:

- EAs are derivative free methods for design optimizations.
- Unlike conventional local search methods, EAs can optimize many variables simultaneously.
- Due to population based approach, EAs search simultaneously in different parts of the search space.
- EAs can be easily implemented on parallel architectures.
- EAs can be easily interfaced with various EM simulation tools in order to optimize design parameters. The inherent mechanisms of EAs overcome local minima in most cases.

3.2.2 Use of Genetic Algorithms in Microwave Design

Many modifications to the simple binary coded GA have been proposed and used by the researchers for the design of microwave components since last decade. Though it is not possible to summarize all works of GA in microwave designs in this chapter, we have tried to cover them briefly and for the in depth knowledge readers are requested to refer [7–15, 45] and the references therein.

An initial review of microwave designs obtained with GA since 1992 to early 1997 was presented by Weile and Michielssen [8]. They covered electromagnetic design applications in four categories: antennas, stratified medium structures, static devices, and miscellaneous. A review on evolving wire antennas such as Yagi and crooked-wire antenna designs using GA was presented by Linden and Altshuler [9]. Johnson and Rahmat-Samii [7] also presented the use of GA in engineering electromagnetics. They collected the works of GA in applications such as design of microwave absorbers, reduction of array sidelobes, designs of shaped-beam antenna arrays, radar target identification, and broad-band patch antennas. Haupt and Werner [10] also described various electromagnetic designs obtained with GA. They used genetic algorithms mainly for the design of antennas, synthesis of array patterns and optimization of scattering patterns.

Many advancements and modifications of GA have been proposed and they have also been used for electromagnetic designs. Johnson and Rahmat-Samii [11] introduced a technique of combining GA and Method of Moments (MoM) for integrated antenna designs. In this technique, GA optimization was combined with a tailored MoM analysis, which involves removal of rows and columns from the Z-matrix instead of refilling the Z-matrix in each iteration of the GA. Their method was used for the design of wide band and dual band patch antennas.

GA has also been used for wireless communication applications. Hong and Dong [12] have proposed two different GA-based efficient searching approaches and applied them to maximum likelihood decoding and distance spectrum techniques to reduce computational complexity for Multiple-Input Multiple-Output (MIMO) systems. Villegas, *et al.* [13] described Electromagnetic Genetic Optimization (EGO) that combines accuracy of full-wave EM analysis with the robustness and speed of parallel computing GA on the cluster supercomputing platform. The EGO was used to design a dual-band antenna element for wireless communication applications.

GA has also been used for problems like array pattern synthesis, array failure detection, and array failure correction. Yan and Lu [14] presented simple and flexible genetic algorithm for pattern synthesis of antenna array with arbitrary geometric configuration. Their approach presented the array excitation weighting vectors as complex number chromosomes and used decimal linear crossover without crossover site. Their method suggested the advantage of avoiding binary coding and decoding, and using simplified approach of chromosome construction. GAs have also been used for the design of oversized waveguide components. Plaum *et al.* [15] optimizes bends for oversized waveguides using GA. In the design of waveguide bends, they optimized curvature function and for corrugated waveguides, the corrugation depth of a bend.

Despite of GA's success and wide use in finding optimum designs, it consumes huge computational time to reach an optimum solution. Moreover due to stochastic nature, sometimes GA may converge prematurely leading to local optima, especially, when the search space is huge and highly nonlinear.

In next subsection, we present a review of works performed for design applications in microwave domain using swarm intelligence based methods and their modifications.

3.3   Microwave Design Using Swarm Intelligence Based Algorithms

3.3.1   Review of Microwave Design Using Particle Swarm Optimization

The use of PSO for electromagnetic & microwave design applications was initially justified in [46, 47]. Rahmat-Samii [46, 48–50] has studied and applied PSO and its modifications to many electromagnetic design applications, especially for the design of various antenna structures and arrays. Robinson and Rahmat-Samii [46] introduced conceptual overview of PSO for electromagnetic community. They indicated the use of invisible wall technique over absorbing and reflecting wall techniques for applying boundary conditions. They showed the use of PSO to the optimization of profiled corrugated horn antenna. Recently, Jin and Rahmat-Samii [48] presented a review on PSO for antenna designs. They illustrated the effectiveness of applying swarm intelligence to design antennas with desired frequency response and radiation characteristics for practical EM applications. They demonstrated the flexibility of PSO to handle both binary and real parameters, and in solving multi-objective problems by applying it to three design problems: design of dual-band patch antenna, artificial ground plane of surface wave antenna, and low-sidelobe antenna array, respectively.

In other modifications to PSO, Ciuprina, et al. [47] presented Intelligent PSO (IPSO) that offered more intelligence to particles by using concepts of group experiences, unpleasant memories, local landscape models based on virtual neighbors, and memetic replication of successful behavior parameters. They tested IPSO on a test function and on Loney's solenoid. Wang et al. [51] presented a combined approach of PSO and Finite-Element Method (FEM) for the design of compact planner microwave filter. They suggested the use of PSO-FEM approach to be useful in wide range of novel filter design. In an another modification to PSO, Mikki and Kishk [52] presented a new physical formalism of PSO technique based on quantum mechanics. They applied this newly developed PSO, known as Quantum PSO (QPSO), to electromagnetic problems such as synthesis of antenna array, and finding equivalent circuit model for dielectric resonator antenna that predicts parameters like Q-factor. The authors proved this algorithm to outperform improved version of classical PSO in convergence speed as well as in obtaining better solution.

One of the drawbacks in performing microwave designs with PSO, due to its iterative nature, is that the overall design process becomes computationally intensive and time consuming. To reduce overall computation time of the design process, PSO can be implemented on parallel clusters. As each particle in PSO acts as an independent agent, it is an inherent characteristic of the algorithm that enables it to be parallelized easily. Jin and Rahmat-Samii [49] presented a method combining PSO and Finite Difference Time Domain (FDTD) method for the design of multi-band and wide-band antennas. They also implemented this method on a parallel cluster to reduce the computational time introduced by full-wave analysis of FDTD method.

It is also observed that many real-time problems have more than one objective. In this case, it is desired to find a solution that optimally balances the trade-off between multiple objectives. Similar to GA, multi-objective optimization is possible to implement with PSO. Xu and Rahmat-Samii [50] shows the use of Multi-Objective PSO (MOPSO) by applying it to two electromagnetic problems: synthesis of 16-element antenna array which is optimized for trade-off between beam efficiency and half-power beam width, and optimization of shape reflector antenna for high gains of multiple feeds.

Researchers have also compared the concepts and performances of PSO with GA. The conceptual difference between PSO and GA is described by Kennedy and Eberhart [53]. According to them, PSO lie between GA and evolutionary computation. The adjustment of particle towards its individual best, and towards global best is conceptually similar to crossover operation used in GA. The benefit with PSO is that it converges in less number of iterations than GA, and requires few parameter settings. PSO has been demonstrated in certain instances to outperform GA [53]. A comparison between PSO and GA on test problems is carried out by Hassan et al. [54]. Boeringer and Werner [55] compared PSO and GA for phased array synthesis problem. They obtained good performance with both the methods. According to them the cost function budget for electromagnetic optimization dominates, and book-keeping requirement for both the algorithms becomes negligible. They also found a simpler implementation and reduced book-keeping appeal of PSO. Despite advantages of PSO such as faster convergence, simple approach, and reduced book-keeping over GA, it may also lead to premature convergence and local minima similar to GA.

3.3.2 Review of Microwave Design Using Bacterial Foraging Optimization

In the literature, very few works showing the use of BFO technique for the design of microwave problems have been reported. However, the works reported in literature [56–58] show that BFO has been used mainly for antenna and array designs. Gollapudi, et al. [56] used bacterial foraging optimization technique for calculating resonant frequency for rectangular microstrip antenna of arbitrary dimension and substrate thickness. They also determined the feed point of the microstrip patch antenna using the same technique. Datta, et al. [57] presented an improved adaptive approach to bacterial foraging algorithm and used it for optimizing both the amplitude and phase of the weights of a linear array antenna, for maximum array factor at any direction and nulls in specific direction. They used principle of adaptive delta modulation to make the algorithm adaptive. Guney and Basbug [58] used bacterial foraging algorithm to achieve null steering in radiation pattern of a linear antenna array by controlling only the element amplitudes. BFO and its improvements have also been used in few other design applications such as job shop scheduling [59], stock indices prediction [60], and optimizing multivariate PID (Proportional-Integral-Derivative) controller [61]. A comparison of PSO and a modified BFO has been shown by [62]. They used both of these SI based algorithms for the critical design applications of millimeter-wave domain.

3.4   Microwave Modeling Using Soft Computing

An inherent and important part of the design process is modeling. Conventional method suggests the use of EM simulation tools for modeling and analysis of microwave components. A typical design problem in EM simulator sometimes takes much computation time even on up-to-date processors. Various regression techniques such as neural networks, response surface methods, kriging, and regression splines can be used as metamodels [63]. A *metamodel* is defined as 'model of the model'. The advantage of these methods is that they respond very fast, and sometimes it is also possible to design components when its closed-form formulas are not available. In this section, we present conceptual overview and a brief review of EM modeling using two soft computing methods namely ANN and SVM. ANN has been widely used for EM modeling, while the SVM is relatively new and is not explored as effectively by EM researchers.

3.4.1   ANN Based Microwave Modeling

A thorough review of ANN applications in microwave CAD is presented by Burrascano, *et al.* [64]. The authors there showed the role of ANN in replacing most CPU intensive part of microwave CAD, namely, yield optimization, tolerance analysis, and manufacturing-oriented design. They illustrated few significant applications, and presented issues for practical implementation. They also introduced self-organizing maps for enhancing model accuracy and applicability. Neural network has been studied and applied widely for the design of RF and microwave components by Zhang and Gupta [6]. After discussing neural network structures and training methods, they provided a general methodology for the development of accurate and efficient Electromagnetically-trained Neural Network (EM-NN) models for use in microwave CAD. They showed ANN-based modeling for various RF and microwave components such as transmission line structures, active devices, microwave circuits, antennas, and systems. At last they described an exciting method - knowledge based neural network by combing microwave knowledge with neural networks and showed its use in RF and microwave design.

Various modifications were performed to simple multilayer perceptron type neural network by researchers according to EM design requirements. Wang and Zhang [16] developed a novel neural network structure, namely, Knowledge-Based Neural Network (KBNN) by combining microwave experience and learning power of neural network. They also developed new error backpropagation training scheme utilizing gradient based $l2$ optimization. They applied KBNN to different microwave modeling problems like circuit waveform modeling, transmission line modeling and MESFET modeling problems, and proved that KBNN gives less testing errors than multilayer perceptrons and it is also efficient when training data is insufficient. Marinova, et al. [17] presented a model by employing neural network inverse algorithm and two feed forward neural networks for solving electromagnetic design problems. The model was applied for the design of magnetic simulation coil and gradient coil. An ANN based approach for modeling of linear and nonlinear circuits was presented by Suntives, et al. [18]. They described a modified hybrid approach for computing S-parameters of microstrip discontinuities based on equivalent circuit extraction and ANN. Ding, et al. [19] presented ANN approach to EM based modeling and optimization in frequency and time domains and used them to nonlinear circuit optimization problems. They presented EM-based time domain neural modeling approach combining available knowledge of equivalent circuits with state-space equations and ANN. Recently, a state-of-the-art review on microwave filter modeling, optimization, and design using ANN techniques is presented by Kabir, et al. [20]. The review included application of ANN on different types of filters such as waveguide cavity filters, simple lower order filter, waveguide dual-mode simple elliptic filters, coupled microstrip line filters, and microwave filters on PBG structure. They demonstrated through results that ANN techniques can produce fast and accurate results and can reduce computational cost compared to conventional and time consuming EM simulations.

Despite their wide use, ANN has also certain drawbacks. It is difficult to find a structure of neural network which includes number of layers, number of nodes in each layer and transfer mapping functions at each layer. There is no general rule to set parameters such as learning rate, momentum, and to find number of training samples for desired accuracy. Although neural networks are capable of achieving high degree of training accuracy in approximating the underlying design process, their generalization ability (error in predicting data not present in training set) is not as accurate. The reason is that neural network tries to minimize the empirical risk (error on training data). Moreover, neural network serves as a black box (i.e., it does not answer how a particular output is obtained). A promising alternative to neural network, which overcomes many of its drawbacks is SVM.

3.4.2 Support Vector Machine Based Microwave Modeling

Though very few works have been reported in literature, we present here a brief review on microwave modeling using SVM. In [65], Angiulli, *et al.* discussed the use of SVR for modeling microwave devices and antennas. They reported that SVR-based model gives better prediction accuracy in less computational time compared to ANN-based modeling approach. Angiulli, *et al.* [66] also used SVR-based approach for electromagnetic inverse scattering problem. Wu, *et al.* [67] used SVR for extracting electromagnetic parameters such as complex permittivity and permeability for magnetic thin film materials. Güneş, *et al.* [68] adapted SVR to the analysis and synthesis of microstrip lines on all isotropic/anisotropic dielectric materials. They found SVR superior to ANN for regression applications due to its higher approximation capability and much faster convergence with sparse solution technique. Güneş, *et al.* [69] also developed SVM model for small-signal and noise behavior of microwave transistor and compared it with ANN model. Martínez-Ramón and Christodoulou [70] introduced a set of novel techniques based on SVM, and applied them to antenna array processing and other problems in electromagnetics. Particularly, SVM was used for linear and nonlinear beam forming, parameter design for arrays and estimating the direction of arrival problems. Modifications to SVR and combination of electromagnetic analysis with SVR have also been tried by some of the researchers. Xu, *et al.* [71] presented an approach for modeling microwave devices based on combination of conventional equivalent circuit model and SVR. They found this approach to be fast and accurate for developing model of SiC MESFET.

Despite SVR's smaller errors and superior generalization capabilities, there are certain challenges in using SVR for microwave design. The accuracy of prediction in SVR depends on selection of hyperparameters. These include deciding values of penalty trade-off parameter C, kernel function and its parameter, and parameter ε of regression tube. Few methods have been suggested in literature but none is guaranteed to give best selection with minimum computational expenses. Moreover, simulation response of microwave components is very sensitive to small changes in its design parameters. So it is challenging for microwave designers and researchers to develop effective models using SVM. Microwave designers may have expert domain knowledge in addition to data sets. Inclusion of this domain knowledge may lead to higher accuracy of the components.

3.5 Microwave Design Using Hybrid Soft Computing Methods

All the individual soft computing techniques are powerful and contribute in the design process, but each of them also have certain drawbacks. ANN and

SVM are efficient modeling techniques but they are not in much use for optimization purpose. GA and PSO are useful in optimization but they require interface with EM full-wave analysis codes. Moreover, due to their iterative nature they are computationally intensive and time consuming. Researchers have also tried to develop hybridization of different soft computing methods by combining two or more methods by removing drawbacks and using advantages of each method [72–75]. This section reviews few works on hybridization of two or more soft computing techniques used for the microwave design applications.

Yang, *et al.* [76] presented a hybrid approach combining PSO with Least-Square SVM (LS-SVM) to improve computational efficiency of FDTD (Finite-Difference Time Domain) method. In this approach, PSO was used to optimize hyperparameters of LS-SVM. Researchers have also tried to combine GA and PSO algorithms exploiting the advantage of both algorithms. A very simple hybridization of GA and PSO was investigated by Robinson, *et al.* [77] and it was tested for profiled corrugated horn antenna. After investigating both algorithms individually, they tested two combinations: GA followed by PSO and PSO followed by GA in which the result of previous algorithm was used as starting point for the later algorithm. According to their results PSO-GA hybrid combination returned the best horn. Gandelli, *et al.* [78] presented a hybrid evolutionary algorithm - Genetic Swarm Optimization (GSO) - by integrating main features of GA and PSO. After performing preliminary studies, GSO was applied to the design of planer reflectarry antenna and found reliable and effective for applications in electromagnetics. Researchers have also developed hybridization of bacterial foraging optimization with GA and PSO, but they have not been investigated for microwave design applications. A hybrid approach combining PSO with BFO was presented and used to optimize multi-modal and high dimensional benchmark functions in [79]. The hybrid method was found to be better than standard BFO and comparable to PSO and its variations on benchmark functions. Kim, *et al.* [80] presented another hybrid approach combining GA and PSO and demonstrated it for solving optimization benchmark problems. They also successfully used it for tuning PID controllers of automatic voltage regulator.

4

Design of Microwave Filters Using Modified PSO

4.1 Overview

Since last decade due to development of electromagnetic simulators by different developers, it has become a general practice to design microwave/millimeter-wave components/systems using EM simulation tools. In this process, an initial design is prepared with coarse and approximate analytical and circuit models. This initial design is farther perfected to obtain a final design by making use of suitable EM simulators. However, this procedure is still semi-automated (where EM analysis is obtained with the help of tools, but the adjustment of parameters according to the response is done manually), tedious, time-consuming and chances of getting local minima are very high.

EM researchers are using various evolutionary and, in recent years, swarm intelligence based algorithms such as genetic algorithm, and particle swarm optimization for the optimization of design parameters [7, 46]. The advantage of this technique is that it reduces manual and laborious task of getting desired response with EM simulations, but there are certain drawbacks of this method. The major drawback in using evolutionary algorithms is that they require large number of iterations, and invoking EM simulation tools in their iterative loops makes the design process computationally expensive and still time consuming. Moreover, increasing design parameters or widening the search range makes them sometimes difficult to converge to global optimum.

Researchers have been working continuously on particle swarm optimization for the improvement of its convergence speed and accuracy [81, 30, 32]. In this chapter, a novel concept of multiple subswarms in PSO algorithm is presented with a hope that searching the feature space in a distributed manner may lead to faster convergence towards global optima. This motivates us to apply the above mentioned approach to complex microwave design process that is computationally expensive and time consuming. The presented algorithm is tested on five benchmark functions first. At last, the applicability of the algorithm has been shown for the design of a specific microwave filter [82]. Though the algorithm presented here can be applied for the design of a variety of microwave components, a microwave filter is considered as a case study.

N. Chauhan, M. Kartikeyan, and A. Mittal: Soft Computing Methods, SCI 392, pp. 35–47.
springerlink.com © Springer-Verlag Berlin Heidelberg 2012

In the next section, a modified PSO considering multiple subswarms is presented and it is tested on both the variations of PSO – Inertia Weight Method (IWM), and Constriction Factor Method (CFM) – as described in variations of PSO in subsection 2.3.1. The resultant modified PSO algorithm (after adapting above modifications) is tested with five benchmark functions commonly used in PSO literatures. The results of modified PSO with above two variations (IWM and CFM) are compared with standard PSO with same variations respectively.

The basic concepts of standard PSO algorithm, and its two basic variations are discussed in subsection 2.3.1. The organization of this chapter is as follows. In section 4.2, a modified PSO algorithm with multiple subswarms is presented and tested with benchmark mathematical optimization functions. Section 4.3 demonstrates the application of the modified PSO algorithm for the design of a specific microwave filter. Finally, section 4.4 presents concluding remarks.

4.2   Particle Swarm Optimizer with Multiple Subswarms

In this section, a new implementation of PSO algorithm considering multiple subswarms is presented [82]. This modification is inspired from human knowledge acquisition process. It can be realized in a society, that an individual obtains information of a particular thing or event based on his own experience (which is stored in his own memory) and the information available from global information sources such as TV, internet, newspaper, etc. These two information sources can be correlated with the local best and global best positions utilized by the particles. However, we (humans) also gain information from our local environment (i.e., from the persons with whom we are in contact). This specifies that every individual is also part of a local group from which he/she also gains information. Inspired by this aspect of an individual in the society, a new modification to PSO algorithm is presented in which a swarm is divided into multiple subswarms. Each particle also considers best information available to the subswarm (local group) in which it belongs. The velocity update equation in the presented method is modified to consider the effects of multiple subswarms. The modified PSO algorithm, which we call PSO with Multiple Subswarms (PSO-MS), is shown in Fig. 4.1.

In the proposed algorithm, first we initialize number of subswarms. The number of particles in each subswarm is defined by total number of particles in the original swarm divided by number of subswarms. In every iteration of the modified algorithm, we also find the subswarm's best (P_{gs}), in addition to the swarm's global best (P_g). To consider the effect of subswarm's best, the velocity update Eq. (2.1) is modified as,

$$v_{id} = v_{id} + c_1 r_1 (p_{id} - x_{id}) + c_2' r_2 (p_{gd} - x_{id}) + c_3' r_3 (p_{gsd} - x_{id}). \qquad (4.1)$$

Initialize number of subswarms
Initialize positions and velocities of all particles in each subswarm randomly
Repeat
    For each subswarm
        For each particle in the subswarm
            Calculate the fitness value $f(X_i)$
            If $f(X_i) < f(P_i)$ then $P_i = X_i$
        End for
        Update P_{gs}, if the best particle in the subswarm has lower $f(X)$ than $f(P_{gs})$
    End for
    Update P_g, if the best particle in the swarm has lower $f(X)$ than $f(P_g)$
    For each subswarm
        For each particle in the subswarm
            r_1=random(), r_2=random(), r_3=random()
            Calculate particle velocity according to equation (4.1)
            Restrict the velocity of particles by $[-V_{max}, V_{max}]$
            Update particle's position according to equation (2.2)
        End for
    End for
Until maximum iteration or minimum error criteria is attained

---

Fig. 4.1 Pseudocode of PSO with Multiple Subswarms (PSO-MS)[82]

Here, the cognitive component is not changed, and the c_1 still represents cognitive coefficient as in standard PSO. We divide social component into two parts: one representing the effect of swarm's global best (P_g), while the other representing the effect of subswarm best (Pgs). The social coefficient c_2 is modified to c_2' and a new coefficient c_3' is added to the equation along with a new random number r_3. Here, we suggest to select the values of c_1, c_2' and c_3' in such a way that $c_1 = c_2' + c_3'$ and $c_2' = c_3'$. These restrictions have been imposed in order to keep the trust of particles equal on cognitive and social components. By keeping equal values of c_2' and c_3' will lead particle equally towards its own global and subswarm best positions.

The modification presented also holds true if the subswarm is not divided into multiple subswarms. If we consider the number of subswarms to be one, then P_g and P_{gs} will produce the same values and the velocity update Eq. (4.1) will reduce to Eq. (2.1) of standard PSO algorithm, provided the value of r_3 is same as the value of r_2 and the values of c_2' and c_3' are chosen as suggested above. In this case, the algorithm of Fig. 4.1 will reduce to standard PSO algorithm as shown in Fig. 2.1. In this work, we test our algorithm (Fig. 4.1) with number of subswarms varying from one to five.

The concepts of IWM and CFM are applied to velocity update Eq. (4.1) in similar way as they were defined in Eq. (2.3) and (2.4) respectively. The values of c_2' and c_3' are selected following the restrictions described earlier. Useful

suggestions for parameter selection for both these variations are obtained
from [83] and [84] respectively. Finally, the positions of particles are updated
similar to standard PSO algorithm according to Eq. (2.2).

4.2.1 Experiment with Benchmark Functions

Benchmark Functions

To compare the performance of proposed algorithm with standard PSO algorithm, five nonlinear benchmark functions commonly found in the PSO
literature [81, 30, 84] are employed. These benchmark functions are shown
in Table 4.1. Among these functions, De Jongs' sphere function and Rosenbrock's functions are unimodal, while rest of the functions are multimodal.
To test these functions the fitness value was taken equal to the function value.

Experimental Settings

In addition to the selection of benchmark functions, there are many experimental settings to be determined such as the swarm size, maximum number
of iterations for each run, the search space size, etc. To keep these settings as
standard as possible, we follow the simulation settings given in [81] whenever
possible. However, the swarm size of 30 is considered for all experiments on
test functions in this work. In order to check the performance with increasing
complexities, the simulations were performed on these functions with three
dimensions 10, 20, and 30 except for Schaffer's f6 function which is of two
dimensions. Maximum iterations are kept at 1000 for Schaffer's f6 function.
Maximum iterations for sphere function are 1000, 2000, and 3000 for dimensions 10, 20 and 30 respectively. For rest three benchmark functions maximum
iterations are 3000, 4000, and 5000 for dimensions 10, 20 and 30 respectively.
The maximum tolerance limit for Schaffer's f6 function is 0.00001, while for
rest of the functions tolerance limit is considered to be 0.01. The range of
search and initializations for test functions are shown in Table 4.2. As the
optimum objective function value for all benchmark functions are near origin,
initializing particles uniformly random in the search range would allow the
particles to be distributed around origin. So to avoid such biasing, asymmetric initializations as suggested in [81] and [85] are used in the experiments.
The maximum velocity V_{max} is considered equal to half the range of search
i.e., $V_{max} = \gamma X_{max}$ [84] where the value of γ is considered to be 1 (remember
the search range is $[-X_{max}, X_{max}]$). Eberhart and Shi [33] also suggest lim-
iting the maximum velocity V_{max} to dynamic range of the variable X_{max} on
each dimension.

Experimental Results and Discussions

In this section, comparison of modified PSO (PSO-MS) is made with standard PSO for both the variations IWM and CFM respectively. Table 4.3

Table 4.1 Benchmark functions [81] (© 2005 IEEE)

Function name	Function
De Jongs's sphere function	$f_1 = \sum_{i=1}^{n} x_i^2$
Rosenbrock function	$f_2 = \sum_{i=1}^{n-1} [100(x_{i+1} - x_i^2)^2 + (x_i - 1)^2]$
Generalized Rastrigrin function	$f_3 = \sum_{i=1}^{n} [x_i^2 - 10\cos(2\pi x_i) + 10]$
Generalized Griewank function	$f_4 = \frac{1}{4000} \sum_{i=1}^{n} x_i^2 - \prod_{i=1}^{n} \cos\left(\frac{x_i}{\sqrt{i}}\right) + 1$
Schaffer's f6 function	$f_5 = 0.5 + \dfrac{(\sin\sqrt{x_1^2 + x_2^2})^2 - 0.5}{(0.1 + 0.001(x_1^2 + x_2^2))^2}$

Table 4.2 Initialization range, search range, and error tolerance [81] (© 2005 IEEE)

Test function	Range of search	Range of initializations
f_1	$[-100, 100]^n$	$[50, 100]^n$
f_2	$[-100, 100]^n$	$[50, 100]^n$
f_3	$[-10, 10]^n$	$[2.56, 5.12]^n$
f_4	$[-600, 600]^n$	$[300, 600]^n$
f_5	$[-100, 100]^n$	$[15, 30]^n$

shows the performance of PSO-MS and PSO with IWM for all five bench-mark functions, while Table 4.4 shows the performance of PSO-MS and PSO with CFM. To neutralize the randomized effects due to probabilistic nature of EAs, the results presented in the tables are average of 50 simulations (runs) for each function. In the tables, when the number of subswarms is one, no division of particles in the swarm takes place and it refers to standard PSO, while the number of subswarms two to five refer to PSO-MS. In the tables, results are shown considering three performance metrics [81]: average achieved optimum value out of 50 runs, number of successful runs out of 50 runs, and average generation of success from the set of successful runs. The best results, considering all three metrics in all cases, are highlighted with bold faces in both the tables.

It is observed from Table 4.3 and Table 4.4 that PSO-MS with two sub-swarms outperforms standard PSO in terms of number of successes and av-erage obtained optimum value. Moreover, it is observed from Table 4.3 and Table 4.4 (in most cases) that PSO-MS with two subswarms obtains similar or better performance in less number of iterations. This leads to reduction in computational expenses. It is observed from both the tables that average generation of success is lower for PSO-MS than PSO for all subswarms. It is also observed for both PSO variations that for generalized Rastrigrin's func-tion with dimension 20 and 30 and Rosenbrock's function with dimension 10, PSO-MS obtains higher average optimum value than PSO. The comparison of Table 4.3 and Table 4.4 shows that for most of the cases PSO-MS and

PSO with CFM obtain better performance in less number of iterations than PSO-MS and PSO with IWM respectively which is also a conclusion in [33]. Finally, a comparison of total number of successes for IWM and CFM shows that PSO-MS with two subswarms obtains more number of successes than standard PSO in both the variations. Here, we also observe the classic case of time-accuracy trade-off. It can be seen from the tables that as the number of subswarms increases, keeping the swarm size same, faster convergence is obtained while decreasing the number of successes. Now, if we need greater accuracy, the swarm size should be increased accordingly, which again results in larger computational time due to more Function Evaluations (FEs) that have to be performed. Hence, in our experiments, two number of subswarms are found to give optimum performance in most of the test cases. Here, the values of c_1, c_2' and c_3' were considered to be 2, 1 and 1 respectively for IWM. For CFM, it was observed that performance is sensitive to the values of c_1, c_2', and c_3' (the similar conclusion is also drawn empirically in [84]). Therefore values of these parameters were manually optimized in CFM for each test case.

4.3 Design of Coupled Microstrip Line Band Pass Filter

In this section, as a case study of microwave components, we have considered the design of microstrip filters. Microstrip filters have become popular due to their small size, low cost, and good performance. Various topologies available for implementing microstrip filters are end coupled, parallel coupled, hairpin, interdigital and combined filters [86–90]. Parallel coupled microstrip lines have been selected for implementation of microstrip filters in this experiment. We have considered the design of Chebyshev band pass filter for approximately 1 GHz bandwidth centered at 2.5 GHz frequency. The specifications for the problem are given in Table 4.5.

The general layout of the coupled microstrip line filter considered for the problem is shown in Fig. 4.2. It is made of cascaded coupled line sections. Skew-mirror symmetry has been used in developing the structure (i.e.,

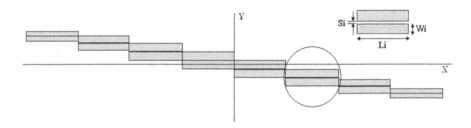

Fig. 4.2 Geometry of coupled microstrip line band pass filter [87]

Table 4.3 Performance comparison of PSO-MS and PSO with IWM [82]

Fn.	Dim.	Average achieved optimum value (Standard deviation)					Number of successes (Average generation of success)				
		Number of subswarms					Number of subswarms				
		1	2	3	4	5	1	2	3	4	5
f_1	10	0.01(−)	**0.01(−)**	0.01(−)	0.01(−)	0.01(−)	50(566)	**50(391)**	50(403)	50(419)	50(428)
	20	0.01(−)	**0.01(−)**	0.01(−)	0.01(−)	0.01(−)	50(1291)	**50(851)**	50(878)	50(915)	50(934)
	30	0.01(−)	**0.01(−)**	0.01(−)	0.01(−)	0.01(−)	50(2081)	**50(1344)**	50(1394)	50(1453)	50(1479)
f_2	10	25.19 (62.34)	**32.85 (84.40)**	55.48 (135.11)	66.78 (122.03)	69.06 (139.94)	1 (2365)	**2 (1997)**	1 (1620)	0 (−)	0 (−)
	20	131.94 (269.92)	56.05 (83.92)	**49.75 (72.59)**	121.24 (210.96)	167.31 (301.86)	0 (−)	0 (−)	**0 (−)**	0 (−)	0 (−)
	30	186.85 (370.73)	**89.27 (141.64)**	103.99 (16.43)	252.12 (394.40)	241.79 (388.54)	0 (−)	**0 (−)**	0 (−)	0 (−)	0 (−)
f_3	10	2.30 (1.55)	**2.20 (1.18)**	2.54 (1.27)	2.74 (1.82)	2.84 (1.44)	3 (1904)	**4 (1349)**	4 (1531)	3 (1943)	2 (1958)
	20	**14.20 (4.68)**	15.79 (5.98)	16.35 (5.82)	15.62 (5.20)	13.69 (4.76)	0 (−)	0 (−)	0 (−)	0 (−)	0 (−)
	30	**31.44 (7.15)**	35.32 (10.23)	37.70 (9.06)	39.36 (9.75)	36.01 (8.70)	0 (−)	**0 (−)**	0 (−)	0 (−)	0 (−)
f_4	10	0.075 (0.035)	**0.054 (0.030)**	0.073 (0.042)	0.08 (0.045)	0.075 (0.041)	0 (−)	**2 (1182)**	1 (1092)	1 (1928)	0 (−)
	20	0.026 (0.026)	**0.021 (0.015)**	0.024 (0.021)	0.022 (0.016)	0.034 (0.027)	22 (2339)	**24 (1511)**	22 (1614)	19 (1612)	17 (1646)
	30	0.02 (0.015)	**0.015 (0.011)**	0.018 (0.015)	0.020 (0.018)	0.027 (0.027)	27 (3222)	**30 (2032)**	25 (2076)	27 (2152)	24 (2177)
f_5	2	0.00097 (0.0029)	**0.00078 (0.0026)**	0.0013 (0.0033)	0.0025 (0.0042)	0.0038 (0.0047)	45 (532)	**46 (417)**	43 (482)	37 (453)	30 (490)
	Total successes (Total iterations)						248 (14300)	**258 (11474)**	244 (11090)	237 (10875)	223 (9112)

Table 4.4 Performance comparison of PSO-MS and PSO with CFM [82]

Fn.	Dim.	Average achieved optimum value (Standard deviation)					Number of successes (Average generation of success)				
		Number of subswarms					Number of subswarms				
		1	2	3	4	5	1	2	3	4	5
f_1	10	0.01(–)	**0.01(–)**	0.01(–)	0.01(–)	0.01(–)	50(66)	**50(48)**	50(81)	50(94)	50(182)
	20	0.01(–)	**0.01(–)**	0.01(–)	0.01(–)	0.01(–)	50(221)	**50(132)**	50(283)	50(348)	50(387)
	30	0.01(–)	**0.01(–)**	0.01(–)	0.01(–)	0.01(–)	50(385)	**50(364)**	50(464)	50(663)	50(953)
f_2	10	**22.12** (**48.18**)	27.77 (65.89)	30.04 (72.22)	53.99 (97.27)	25.81 (83.83)	**6** (**2547**)	4 (1968)	1 (2636)	1 (1428)	1 (2435)
	20	43.36 (89.99)	**38.32** (**75.41**)	46.28 (84.60)	70.97 (115.01)	89.17 (121.25)	2 (3116)	**3** (**3266**)	3 (3784)	0 (–)	0 (3930)
	30	54.29 (84.64)	**41.33** (**65.62**)	65.11 (149.2)	148.26 (399.28)	120.20 (199.99)	1 (3712)	**1** (**3331**)	0 (–)	1 (4859)	0 (–)
f_3	10	2.30 (1.48)	**2.32** (**2.05**)	2.38 (2.10)	3.10 (2.05)	4.34 (2.81)	4 (1920)	**7** (**2085**)	5 (2533)	4 (2494)	1 (2760)
	20	11.44 (4.15)	12.07 (5.65)	14.65 (5.33)	15.84 (5.35)	14.78 (7.51)	0 (–)	0 (–)	0 (–)	0 (–)	0 (–)
	30	**26.34** (**8.17**)	30.0 (10.65)	35.04 (11.94)	35.26 (10.18)	38.02 (14.42)	0 (–)	0 (–)	0 (–)	0 (–)	0 (–)
f_4	10	0.069 (0.034)	**0.059** (**0.029**)	0.066 (0.035)	0.073 (0.042)	0.079 (0.051)	1 (1376)	**2** (**894**)	0 (–)	2 (353)	1 (554)
	20	0.032 (0.059)	**0.026** (**0.020**)	0.028 (0.023)	0.030 (0.030)	0.069 (0.083)	18 (263)	**21** (**305**)	19 (350)	20 (399)	8 (453)
	30	0.037 (0.063)	**0.017** (**0.016**)	0.035 (0.039)	0.13 (0.21)	0.44 (2.14)	23 (419)	**30** (**413**)	25 (493)	7 (686)	5 (741)
f_5	2	0.0013 (0.0033)	**0.00039** (**0.0019**)	0.0013 (0.0013)	0.0013 (0.0031)	0.00244 (0.0041)	40 (326)	**47** (**432**)	43 (456)	38 (592)	33 (517)
		Total successes (Total iterations)					245 (14351)	**264** (**13138**)	243 (11094)	223 (11916)	200 (12912)

Table 4.5 Design specifications of coupled microstrip line band pass filter [82]

Center frequency f_0	2.5 GHz
Bandwidth ($\triangle f$)	1.0 GHz
Pass-band ripple (L_a)	0.3 dB
Source and load impedance (Z_0)	50 Ω

structure on the left side of Y-axis is obtained by rotating right side structure 180° about Z-axis). In this problem, for each coupled stripline section, we have considered length of stripline (L_i), width of stripline (W_i), and separation between striplines (S_i) as design parameters as shown in Fig. 4.2. The number of coupled stripline sections is considered to be 4 on either side of vertical axis, thus making total number of design parameters 3*4=12. The range of values considered for the design parameters are shown in Table 4.6. The goal here is to get optimum combinations of these parameters that obtain bandwidth of 1.0 GHz resonanced at 2.5 GHz while minimizing reflection coefficient S_{11} within the band.

Table 4.6 Range of values for the design parameters [82]

Name of design parameter	Range (mm)
Length of stripline (L_i)	[20.5, 23.0]
Separation between striplines (S_i)	[0.1, 0.4]
Width of stripline (W_i)	[1.0, 3.5]

4.3.1 Experiment and Results

The design of filter was carried out using PSO-MS with both variations IWM and CFM considering two subswarms. To compare the output response of PSO-MS, the design optimization was also carried out using standard PSO, again with both the variations IWM and CFM. The design problem considered here has equality constraint of maintaining center frequency at 2.5 GHz. In this design with PSO-MS and PSO, a well-known penalty based approach as described in [29] is adopted for constraint handling. For handling boundary constraints, a three wall method (i.e., reflecting wall, absorbing wall, and invisible wall) is suggested in [46]. In the absorbing wall method, when the particle violates the boundary constraint in any dimension, the velocity in that dimension is made zero. In our experiments, we used a method similar to absorbing wall, but with small variation. In our method (which we refer to it as *sticky wall* method), when the boundary constraint is violated, the velocity is not reduced to zero, but the cut-off limit is imposed to the position of particle in that direction so that particle sticks to the wall and do not cross the range. This method is helpful, particularly, when the optimum value for that dimension is at the boundary. If the optimum is not at the boundary, then the particle will eventually come into the design space again with the effect of individual best and global best positions. The fitness function considered for the design in this problem is given as,

$$FitnessFunction = a_1(|BW - 1.0|) - a_2(nf_{s11}/nf) + a_3(|CF - 2.5|) \quad (4.2)$$

where BW is the bandwidth obtained in GHz (BW is obtained at 10 dB level from the graph of S-parameters), nf_{S11} is number of frequencies within the desired band for which $S_{11} < -15 \ dB$, nf is total number of frequencies considered during simulation in the desired band, CF is the frequency at which resonance is obtained. The weights a_1, a_2, and a_3 are constants and can be chosen as per the requirement of larger BW, lower reflection (S_{11}) or resonance matching. In this experiment the values of a_1 and a_2 were considered to be 1, while the value of a_3 was considered to be 2.

In this work, a MoM based EM simulator IE3D [91] was used to obtain filter response. A typical simulation using this simulator takes few minutes (approx. 3-4 min.) of time to give the response on a computer with core2duo processing capacity. A program interface was prepared and the EM simulator was invoked in iterative loop of the optimization algorithms. As evolutionary algorithms such as PSO require many iterations to converge to a desired objective, the resultant problem becomes computationally expensive if simple PSO is used.

In the design optimization process, PSO-MS (with two subswarms) and PSO were executed with 10 particles for fixed 50 iterations. The convergence graph for all four simulations is presented in Fig. 4.3. It is observed from the graph that PSO-MS converged faster towards better fitness value than PSO for both the variations. This can be observed from the graph that at any time, the value obtained by PSO-MS is closer to the desired objective than the value obtained by PSO in both variations. The optimum filter response (S-parameters) for IWM is shown in Fig. 4.4, while for CFM is shown in Fig. 4.5. The BW of 1.01 GHz and 0.75 GHz were obtained with PSO-MS and PSO respectively for IWM, while the BW of 1.02 GHz and 0.99 GHz was obtained with PSO-MS and PSO respectively for CFM. It can be seen from the figures that for IWM, PSO-MS obtains higher BW and produce the design which is near to the desired objective than PSO with same number of iterations, while in case of CFM, both PSO-MS and PSO obtain desired results.

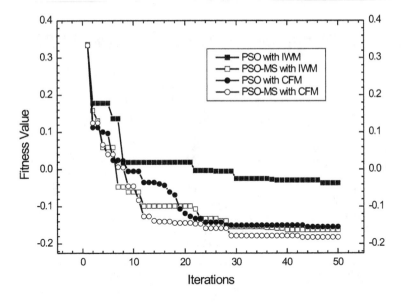

Fig. 4.3 Convergence of PSO-MS and PSO for IWM and CFM [82]

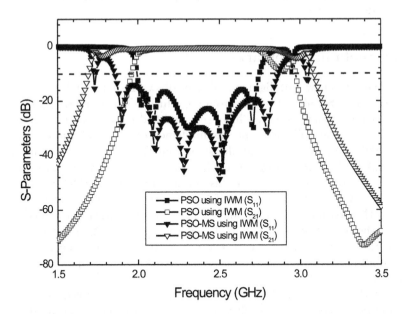

Fig. 4.4 Comparison of filter response using PSO-MS (with two subswarms) and standard PSO with IWM [82]

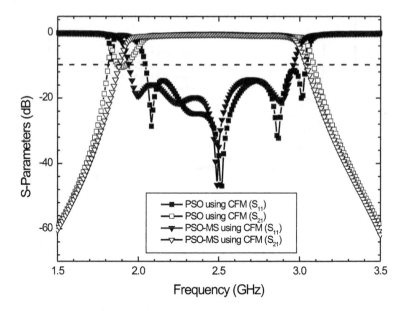

Fig. 4.5 Comparison of filter response using PSO-MS (with two subswarms) and standard PSO with CFM [82]

4.4 Concluding Remarks

In this chapter, a new paradigm of multiple subswarms for searching parameter space with PSO algorithm has been introduced. The social component of PSO's velocity update equation has been modified to consider the effects of multiple subswarms. The concept was implemented for two basic PSO variations - IWM and CFM. The results show that the modified PSO with two subswarms leads faster convergence while improving the quality of solution compared to standard PSO. It can also be observed that PSO-MS gives desired results within the framework of time-accuracy trade-off. We have also observed that parameters c_1, c_2' and c_3' in CFM are critical for the convergence of the algorithm and have to be chosen carefully. All simulations on test functions were carried out with swarm size 30. Yet improved performance can be obtained by using swarm size 50 or greater but it would also increase number of FEs.

At last the modified PSO was used for the design of coupled microstripline band-pass filter and its result was compared with the design results obtained using standard PSO. The comparison of results for the microstrip filter design problem proves the applicability of modified PSO method in real time microwave/millimeter-wave design problems where computational time is an important factor. However, the concept of modified PSO is demonstrated for only one design problem as a case study, it can also be applied to other challenging microwave design problems without change in the algorithmic part. Another advantage of this algorithm is that it can be parallelized easily as each subswarm's computation can be done on a separate processor.

5

Design of Microstrip Antennas and Other Components

5.1 Overview

Since last few decades, commercial growth in the use of microwave products has created necessity for efficient design of components in a small time frame. A Computer Aided Design (CAD) approach is adopted to minimize the time required to obtain an optimized design. In order to use CAD models, the results predicted by them should be consistent with the actual results [92, 93]. Due to very small wavelengths involved in microwave design, it requires high precision during the design. Hence, it is not easy to model components of RF/microwave and millimeter-wave domains.

In the design process, full-wave EM simulators are widely employed for the analysis of microwave components. These simulators give desired solutions and are called *fine models*. One of the limitations of these models is that they are computationally expensive, especially when they are invoked in the optimization algorithms such as GA, PSO, etc. [94]. Besides this, a proper design may not be obtained using EM simulators alone, when the number of constraints are more. In order to remove these drawbacks, we may use mathematical curve fitting techniques that obtain data from experimental measurements (or from EM simulators). These models are referred to as *coarse models* [95]. However, these coarse models have inherent limitations of accuracy and validity over a restricted range of parameter values.

A few curve fitting techniques have been used by mathematicians and researchers from different disciplines to generate coarse models. Microwave researchers have used artificial neural network for the design of RF and Microwave components [6]. ANN models trained by EM simulated data have been used to balance the trade-off between computation time and accuracy since last decade [6, 95, 96]. ANN models, once trained with a data set, have provided models that are almost as accurate as fine models and as fast as coarse models. However, the generalization accuracy achieved by the ANN based models of microwave components needs improvement to increase the effectiveness of CAD. A strong competitor of ANN which has gained popularity due to its generalization ability in recent years is Support Vector Machine (SVM). SVM

N. Chauhan, M. Kartikeyan, and A. Mittal: Soft Computing Methods, SCI 392, pp. 49–67.
springerlink.com   © Springer-Verlag Berlin Heidelberg 2012

is a machine learning tool designed to automatically deal with the accuracy-time trade-off by minimizing an upper bound on the generalization error [38]. In this chapter, we present an SVM based framework for efficient modeling of microwave components. We also present a hybrid approach combining SVM with evolutionary and SI based algorithms such as GA and PSO respectively, and use them for the design of specific microwave components [97, 98]. Three different design problems have been chosen for demonstrating the effectiveness of SVM based modeling and the proposed hybrid approach. The three problems include: effective modeling of a one-port microstrip via, design of a circular polarized microstrip antenna, and design of a simple aperture coupled microstrip antenna.

The chapter is organized as follows. Section 5.2 presents a comparison between ANN with SVM, conceptual background of SVM, and framework for SVM based microwave modeling. The hybrid approach combining SVM with genetic algorithm and with particle swarm optimization respectively is presented in section 5.3. Section 5.4 presents modeling and design of three microwave components mentioned earlier. Finally, concluding remarks is given in section 5.5.

5.2   Microwave Modeling with Support Vector Machine

As we discussed in subsection 2.4.2, SVM can be used for classification as well as regression (function approximation) tasks. The conceptual background of SVM for regression task (which is known as support vector regression (SVR)) was also discribed in the same subsection. In this section, we describe the framework for microwave modeling using SVR as given below.

Identification of input and output: The preliminary step towards developing an SVR model is the identification of inputs and outputs of the problem to be modeled. Inputs for this work are the variable design parameters of the microwave component being modeled. Model outputs are determined based on the purpose of the model. This is typically the metric (e.g. S-parameters, bandwidth, etc.) that is used for evaluating the performance of the component being modeled.

Generation of training and test data: Training data for modeling the component is obtained by performing experiments or simulations for a set of sample inputs. The sample inputs are chosen in such a way that the behavior of the component with respect to design parameters is completely obtained. In our experiments, the initial structures (training data) for simulation are identified based on Design Of Experiments (DOE) methodology [93]. This method is generally used to study and model the input-output relationship for a given process or component. The range of all input parameters of the component being modeled are identified. The range of an input parameter forms a line segment in single dimension, and if all the input parameters are considered simultaneously in N-dimensional space, it would form a multidimensional cube (hypercube) as shown in Fig. 5.1. The initial value of the design structures are chosen as the corners of the hypercube, the midpoint of the edges of the hypercube, and the center of the hypercube. To generate test data, sample inputs are generated randomly from the range of design parameters. The experiments or simulations are carried out on these training and test inputs to get actual training and test outputs.

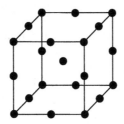

Fig. 5.1 Hypercube illustrating the selection of initial design parameters. Each dimension indicates different design parameter. Length along the edge indicates operating range of the parameter, while the dark circles indicate the initial design parameters that are chosen.[97]

Input scaling: We scale each input parameter value in the train data set between +1 to -1. This form of normalization prevents one of the variables to dominate in the prediction of output variables.

Selection of Kernel function: Various kernel functions such as linear, polynomial, Gaussina RBF, exponential RBF, sigmoid, splines, etc. can be used for mapping of input space to high dimensional feature space and they are elaborated briefly in [63, 42]. The two popular choices for kernel functions are,

$$\text{RBF(Gaussian) Kernel: } K(x, x') = \frac{exp(-\|x - x'\|^2)}{2\sigma^2} \tag{5.1}$$

$$\text{Polynomial Kernel: } K(x, x') = (<x + x'> +c)^p \tag{5.2}$$

where $1/(2\sigma^2) = \gamma$ (gamma) of Gaussian-RBF kernel and p (degree of polynomial kernel) are critical parameters, while c in polynomial kernel is a constant.

Model selection: One of the important choices in developing an SVR model is the selection of model parameters (also known as *hyperparameters*) which include kernel parameters, the penalty of estimation error (C), and the value of ϵ. The value of kernel parameter implicitly defines the structure of the high dimensional feature space where a maximal margin hyperplane is found. Too rich feature space would result in over-fitting of data and if the kernels are too poor, data would not be predictable. Parameter selection thus involves obtaining the optimal values of C, ϵ and σ which would maximize the generalization ability. One of the model selection techniques commonly employed in literature is grid search. The elementary step in grid search is *cross validation* [99]. In a k-fold cross-validation, training set is first divided into k subsets of equal size. Sequentially each subset is tested using the SVR, which is trained on the remaining k-1 subsets. Thus, each instance of the whole training set is predicted once. The cross-validation accuracy is a measure of error in prediction of the data. The goal of model selection is to determine which combination of C, ϵ and σ has the maximum cross validation accuracy (minimum error). Various combinations are tried for the three parameters by sampling the search space at discrete intervals. Once the combination with minimum mean squared error is found, the search is performed around the combination with a reduced sample interval. This procedure is repeated until there is no significant improvement in the cross validation accuracy. Few other methods for tuning hyperparameters of SVM are suggested in literature [76, 100].

Training: Once the optimal parameters for the kernel are chosen, trained SVR model is prepared with training data. This process internally identifies the support vectors in the train data. Support vectors are the input points (part of training data) which are closest to the optimal hyperplane. The output of training is an SVR model. The regression model thus obtained is then used to predict the output values (performance) for various inputs (design parameters). If the accuracy of the model is not within the acceptable limits, the process is repeated with simulations performed using more variations in design parameters. The accuracy is measured using cross validation mean squared error.

Testing: The accuracy of model in predicting unseen data is verified by predicting the performance on an independent test data set. The test data set is also scaled between +1 and -1 using the same scaling parameters as used in the *input scaling* step. The output field of the test data set is predicted using the model file obtained after training SVR. The accuracy of prediction is defined in terms of Mean Squared Error (MSE) and Average Relative Error (ARE). MSE and ARE are computed as follows:

$$Error = PredictedValue - ActualValue, \tag{5.3}$$

$$MSE = \frac{\sum_{i=1}^{n} Error_i^2}{n}, \tag{5.4}$$

$$ARE = \frac{\sum_{i=1}^{n} \left(\frac{Error_i}{ActualValue_i} \right)}{n}, \tag{5.5}$$

where n is the number of sample points in test data set. After computing the accuracy on the test data set, it is determined if the accuracy of the model is acceptable by comparing it with a threshold value. The threshold is chosen based on how much accuracy can be obtained while making a physical design of the actual structure.

Step for improving the accuracy: As SVR is not data intensive compared to ANN, we may start with fewer training samples to model the component. If the generalization accuracy is below the acceptable level, additional training data are included. The dimensions of the additional structures are chosen as the midpoints of the points chosen during first iteration. The modeling process is repeated with additional training data. The need for this iterative process is due to the number of structures that have to be simulated for obtaining an accurate model can not be predetermined for a specific component.

5.3 Modeling of One-Port Microstrip via Using SVR

5.3.1 Problem Description

The first component considered for modeling is a broadband GaAs one-port microstrip via. The top view of the via structure is shown in Fig. 5.2 [6, 95]. It consists of two planes, the ground plane and the substrate plane. The ground plane is placed at a height 0 and the substrate plane is placed at a height H_{sub} above the ground plane. The excitation signal is supplied to the component by means of a feedline through a port in the ground plane. The structure shown in Fig. 5.2 is a metal sheet on top of the substrate. Via is a hole drilled from the metal plate to the ground plane. ε_r is the dielectric constant of the substrate medium. The height of the substrate (H_{sub}), the dielectric constant (ε_r), and all loss parameters are considered constant for this experiment. The width of the incoming microstrip line (W_l), the side of the square shaped via pad (W_p) and the diameter of the via hole (D_{via}) are the three variable geometrical parameters. The component is characterized in terms of its S-parameter (S_{11}) which specifies the proportion of input energy that is reflected back.

While using this component in a microwave circuit, different combinations of the variable design parameters are tested to obtain the optimal performance of the overall circuit. Thus, the objective of the modeling in this problem is to predict the performance of the component for a particular design specification. This is done by obtaining a relationship between particular values of the variable design parameters and the S-parameter corresponding to those particular values. The characterization of the component at various frequencies is also required, since the circuit in which the component is used may be operating at various frequencies. The input variables we consider for modeling are three ratios of geometrical parameters W_l/W_p, W_p/D_{via}, and W_l/H_{sub}, and the frequency is also considered as fourth parameter. The range of these parameters considered for modeling are shown in Table 5.1. Output variable is the magnitude of S_{11} (S-parameter for one port) referenced at 50 Ω port termination.

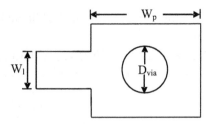

Fig. 5.2 GaAs microstrip ground via geometry [6, 95] (© 1996 IEEE)

Table 5.1 Range of input parameters for modeling of one port GaAs Microstrip Via [6, 95] (© 1996 IEEE)

Input Parameter	Range
Frequency	[5-55] GHz
W_l/W_p	[0.3-1.0]
W_p/D_{via}	[0.2-0.8]
W_l/H_{sub}	[0.1-2.0]

5.3.2 *Experiment and Results*

For this experiment, the training data was obtained by performing simulations on the IE3D simulator [91]. Simulations were performed with frequency varying from 5 to 55 GHz. Each simulation is performed by varying the values of input parameters in the range shown in Table 5.1. In order to minimize the number of simulations that need to be performed as well as capture the complete behavior of the component the initial structures that are to be used for training need to be carefully chosen carefully. The procedure used for selecting the initial training set is discussed in previous section. If it is found that the input-output relationships of the component have not been sufficiently captured (when cross validation error is high), additional simulation points are added to fit the higher order nonlinearities.

After obtaining the training data, the SVR model of the component was obtained using the procedure described in previous section. In order to compare the performance of SVR based model, a Feedforward Neural Network (FF-NN) model was designed. Best results for the feed forward NN were obtained using 10 neurons in the hidden layer. TanSigmoid was used as the activation function of the hidden layer, while a linear function is used for the output layer. The ANN model was designed using MATLAB Neural Network toolbox. Over-fitting of data during training is prevented by using simultaneous testing. The overall data set was divided into train data and

simultaneous test data. Neural network was trained using the train data and
accuracy of training is determined by predicting the simultaneous test data.
The training iteration was repeated as long as the error in predicting simultaneous test data is non-increasing. The training accuracy is also verified on
an independent test data set. Fig. 5.3 illustrate the plots of the actual values
with the values predicted by FF-NN and SVM respectively. The comparison
of S-parameters (magnitude) using SVM, FF-NN and actual values for two
independent test cases are shown in Fig. 5.4 and Fig. 5.5 respectively. The
accuracy of modeling for the various models is given in Table 5.2. It can
be noticed that SVM provides significant improvement in performance over
ANN (FF-NN) model.

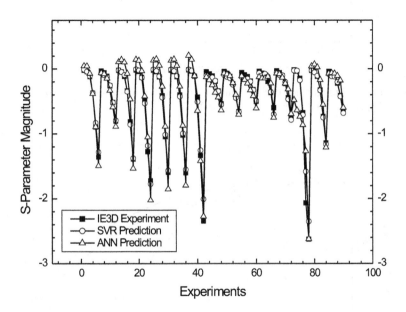

Fig. 5.3 Plot of actual values vs. predicted values using SVR and Feedforward NN
[97]

Table 5.2 Accuracy of modeling one-port microstrip via [97]

Models	MSE	ARE
Feedforward NN	0.0012	0.0290
SVM	5.836e-5	0.005164

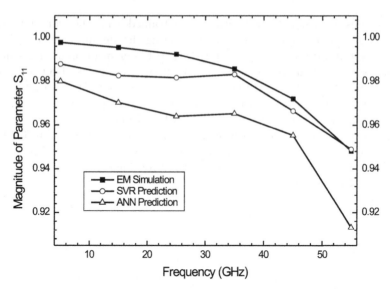

Fig. 5.4 Plot of magnitude of S-parameter against frequency for dimension: W_1/W_p = 0.4, W_p/D_{via} = 0.3 and W_1/H_{sub} = 0.2 [97].

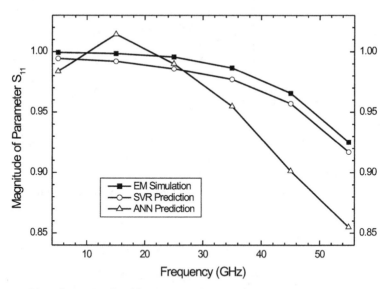

Fig. 5.5 Plot of magnitude of S-parameter against frequency for dimension: W_1/W_p = 0.75047, W_p/D_{via} = 0.34897 and W_1/H_{sub} = 1.0 [97].

5.4 Support Vector Driven GA/PSO Algorithm: A Hybrid Approach

In recent years, the EM analysis of most microwave components is performed using EM simulation tools. Evolutionary algorithms (EAs) such as genetic algorithm and Swarm Intelligence (SI) based algorithm such as particle swarm optimization can be used for optimizing design parameters of microwave components. This requires EM simulation tools to be invoked in the optimizing loop of EA/SI algorithms. Due to iterative nature of these algorithms, the entire process becomes computationally expensive. A modified method with PSO was discussed in chapter 4. It reduced the number of iterations up to some extent in order to reduce the computational expenses. An alternative way of solving this problem is to make use of *metamodel*. A metamodel is a 'model of the model'. The number of metamodeling techniques such as ANN, SVR, response surface methodology, regression splines, etc., [63] can be used to a create model of the time consuming EM simulation process. Since SVM has proved to be successful over ANN model for many applications, we use support vector regression method for creating metamodel (model henceforth) of the process.

In this section, we present a hybrid approach combining SVR with EAs such as GA and PSO for the design of microwave components [97, 98]. We call this approach as Support Vector driven Evolutionary Algorithms (SVEA) when GA is used as optimizer and as SVPSO when PSO is used as optimizer. In this method, we first create SVR model for getting response of the component to be designed. The data for training SVR can be obtained using experiments with EM simulators by varying design parameters in a pre-decided range. The range of design parameters are decided based on the simulation response of initial geometry and expert domain knowledge. The data generated empirically are used to create SVR model following steps specified in section 5.2.

The SVR model, thus created, is used as a metamodel replacing the complicated and time consuming parametric analytical procedure carried out by the EM simulator. The SVR model is invoked in the optimization loop of GAs as a fitness function for optimizing design parameters of microwave components. The exciting advantage obtained with SVR model is that it responds quickly (approximately in milliseconds) compared to iterative parametric analysis of EM simulator response (approximately in minutes and generally it depends on the complexity of the structure). Here, we use two optimizers - GA and PSO to optimize the design parameters by invoking SVR model as its fitness function. As SVR model responds quickly, it is easy for to perform large number of iterations to optimize design parameters. A simple block diagram for the proposed approach along with conventional approach of using EM simulator, and the design approach using GA/PSO (which invokes EM simulator directly as its fitness function) is shown in Fig. 5.4. It should be noticed here that number of simulations required to generate training data to prepare SVR

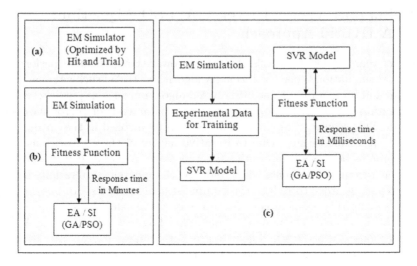

Fig. 5.6 (a) Traditional approach of design using EM simulator (Using expert domain knowledge, and hit-and-trial method) (b) Design using GA/PSO by directly invoking EM simulator as its fitness function (c) Design using proposed approach SVGA/SVPSO [98]

model are far less than number of simulations required if the optimization would have conducted by directly invoking the EM simulation tools in the fitness function of optimizers (see (b) of Fig. 5.4).

In the next section, we present three experiments demonstrating the effectiveness of SVR based modeling and efficiency of the presented hybrid approach in designing microwave components.

5.5 Experiment

5.5.1 Design of Circularly Polarized Microstrip Antenna Using Support Vector Driven Genetic Algorithm

Problem Description

Aperture coupled Microstrip Antennas (MSA) are widely used in wireless applications [88, 101–106]. The circular polarization of antenna is required to make the devices independent of orientations. A method for improving axial ratio bandwidth of circular polarized microstrip antenna is given in [107]. In this experiment, we used the presented hybrid soft computing approach for improving the bandwidth of the circular polarized microstrip antenna. The geometry of the MSA considered for the design here is shown in Figure 5.7. It consists of two dielectric layers both having same dielectric constants and loss tangents. The antenna design employs air gap between two substrate layers. The top layer is of dielectric sheet which supports patch. The bottom layer is a dielectric sheet which supports microstrip feed line on one side and slot on other side. In the proposed design, antenna patch is perturbed by inserting different length slits (horizontal and vertical, see Fig. 5.7(a)) for the generation of circular polarization [101]. The right circular polarization is obtained here by maintaining length of horizontal slit less than the length of vertical slit [103].

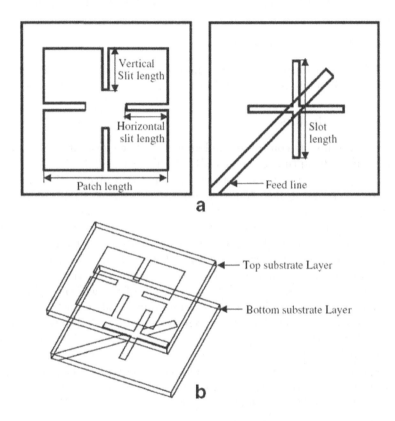

Fig. 5.7 Geometry of the MSA (a) Patch shape and feed arrangement, and (b) 3D view [98]

Our aim in designing Circular Polarized Microstrip Antenna (CPMSA) is to maximize its Axial Ratio (AR) bandwidth (for AR < 3 dB) with resonance at 2.6 GHz. The Axial Ratio Bandwidth (ARBW) of microstrip antenna is given as,

$$ARBW(\%) = (\frac{f_H - f_L}{f_c}) \times 100 \qquad (5.6)$$

where f_H and f_L are upper and lower frequencies considered at 3 dB from
AR vs. frequency plot respectively and f_c is the resonant frequency. The
constraints considered for the optimization are to maintain Voltage Standing
Wave Ratio (VSWR) in its feasible range (A VSWR $\leq$ 2 ensures good performance for wide band operation), to maintain right circular polarization,
and to maintain working frequency at 2.6 GHz. Four design parameters of
microstrip antenna (see Fig. 5.7) namely horizontal slit length, vertical slit
length, patch length (squared patch is considered), and slot length are considered for optimization. The objective function considered for optimization
is defined as,

$$f_{obj} = -a(ARBW) + b|CF - 2.6| + c(VSWR - 2.0) \qquad (5.7)$$

where $ARBW$ is defined in Eq. (5.6), and CF is the resonant frequency. The
coefficients a, b and c are user defined constants.

Experimental Results

Here, two experiments considering different dielectric constants and air gaps
were conducted to optimize the $ARBW$ of CPMSA at 2.6 GHz band. The
material dielectric constants in both experiments were fixed at 7.2 and 2.33
respectively. The air gap in both experiments was kept 3 mm and 4 mm
respectively. The stub length for both experiments were manually optimized
to 9.005 mm and 10.5 mm respectively. Two sets of training data (42 and
75) were generated from IE3D for two experiments respectively. The first
experiment was conducted by considering all four design parameters while
the second experiment was conducted by considering only three design parameters namely length of horizontal slit, length of vertical slit, and patch
length. The parameter slot length was optimized manually prior to the experiment. This was done to reduce the search space and thus improve the model
accuracy. Apart from design parameters, the parameters of support vector
regression such as type of kernel, parameter of the kernels (σ for RBF kernel,
order of polynomial for polynomial kernel), trade-off parameter C, and ε parameter of loss function were optimized manually by measuring their Cross
Validation accuracy (CVacc) and Root Mean Squared Error (RMSE), which
are defined as,

$$CVacc = \frac{c}{t} \times 100 \qquad (5.8)$$

and

$$RMSE = \sqrt{\sum_{i=1}^{n}(y_i - f(X_i))^2}. \qquad (5.9)$$

Here, c indicates number of test data that fall under regression tube (i.e.,
within the desired predefined prediction limit), and t indicates total number of test data. Here, y_i is the experimental value using IE3D and $f(X_i)$

is the predicted value. In the experiment, RBF kernel (with value of $\sigma =$ 0.5, 0.4 respectively for two experiments) was found to give best performance. The optimum values of C and ε are kept ∞ and 0.02 respectively for both the experiments. The Matlab toolbox of SVR [108] was used to implement the algorithm. The 7-fold and 5-fold cross validation was used to measure the performance in both the experiments respectively.

In this example, we used GA for optimizing SVR model of CPMSA, hence the hybrid approach SVGA has been used for this design. To compare the results of proposed approach, similar experiments were conducted for the same datasets using another hybrid approach of Neural Network driven Genetic Algorithm (NNGA). In this method, the empirical approximation model was generated using neural network [35]. Three layer feed forward backpropagation neural network with Levenberg-Marquardt transform function and 0.1 learning rate was used to create model. The number of hidden neurons to create ANN model were 4 and 5 respectively for both the experiments. The predictions using SVR and ANN models are presented in Fig. 5.8 and Fig. 5.9 respectively for both the experiments. The performance of the models is presented in Table 5.3 using both CVacc and the corresponding RSME as defined in Eq. (5.8) and Eq. (5.9) respectively.

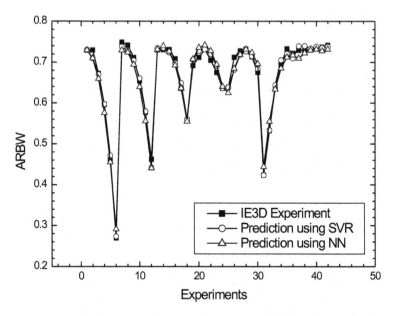

Fig. 5.8 Predictions using SVR model and ANN model with IE3D experiment for Experiment 1 [98]

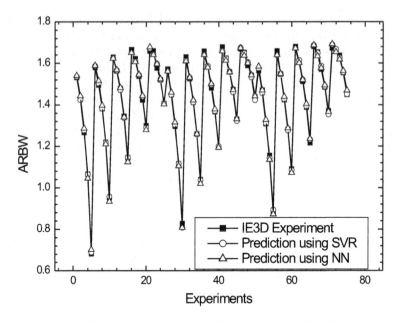

Fig. 5.9 Predictions using SVR model and ANN model with IE3D experiment for Experiment 2 [98]

Table 5.3 Performance of the SVR and NN models [98]

Experiment	SVR Model		NN Model	
	CVacc	RMSE	CVacc	RMSE
Experiment 1(7-fold)	74.8	0.4623	66.67	0.6073
Experiment 2(5-fold)	82.33	0.3583	79.0	0.5587

In this experiment, GA with 50 population size, roulette wheel selection, 0.8 crossover and 0.01 mutation was simulated for 2000 generations to get the optimized values of each experiment. The optimized parameters and optimum ARBW for both the experiments are summarized in Table 5.4. The ARBW was calculated at 3 dB from the axial ratio plot. The ARBW along with optimized design parameters are obtained using EM simulator and are shown in Table 5.4 for the purpose of comparison. The characteristic of AR vs. frequency for both methods are shown in Fig. 5.10 and 5.11 respectively for both the experiments. It is observed from the table that the ARBW obtained with SVGA approach is much closer to the value obtained from experiment with EM simulator compared to NNGA approach in both the experiments.

Table 5.4 Optimized parameters and ARBW using SVGA, and NNGA [98]

Optimized Parameters	Experiment 1 (Dielectric const. = 7.2)		Experiment 2 (Dielectric const. = 2.33)	
	SVGA	NNGA	SVGA	NNGA
H.slit length (mm)	8.726	8.779	10.35	10.27
V.slit length (mm)	9.108	9.117	11.8	11.795
Patch length (mm)	24.47	24.41	31.996	31.985
Slot length (mm)	13.9	13.66	18.5	18.5
% ARBW	0.70	0.73	1.71	1.59
% ARBW(Using EM simulator)	0.75	0.65	1.68	1.68

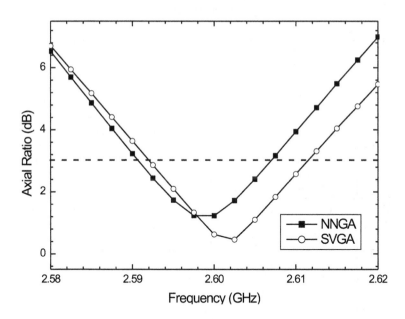

Fig. 5.10 Plots of Axial Ratio vs. Frequency for Experiment 1 [98]

The plots of VSWR vs. frequency are presented in Figure 5.12 and Figure 5.13 for both the experiments respectively. It is observed that the optimized results satisfy minimum VSWR criteria at desired 2.6 GHz frequency.

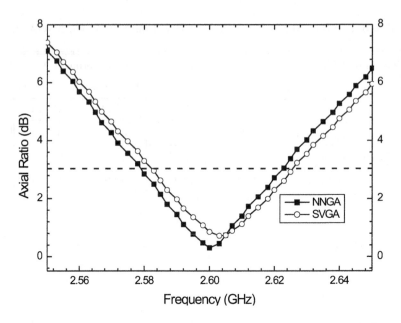

Fig. 5.11 Plots of Axial Ratio vs. Frequency for Experiment 2 [98]

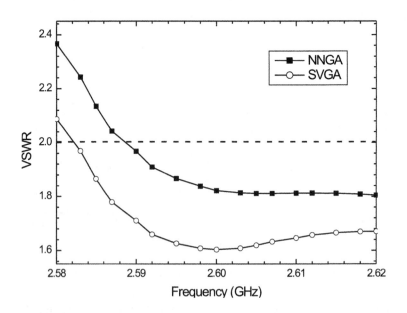

Fig. 5.12 Plots of VSWR vs. Frequency for Experiment 1 [98]

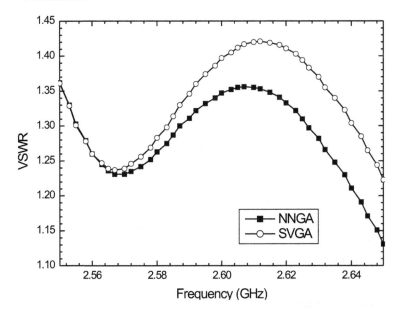

Fig. 5.13 Plots of VSWR vs. Frequency for Experiment 2 [98]

5.5.2 Design of Aperture Coupled Microstrip Antenna Using Support Vector Driven PSO Algorithm

Problem Description

The third structure we considered is a simple aperture coupled microstrip antenna [88, 102, 104, 109]. The geometry of the antenna structure considered for design is shown in Fig. 5.14. Here, two substrate layers are placed on each other without any air gap. The top substrate contains a rectangular patch, while bottom substrate contains slot on one side and feedline on other side. The variable design parameters for the chosen structure are the length of the slot, length and width of the patch, and the length of the open circuited stub. The other parameters such as dielectric materials and their heights were kept constant and they are listed as: $\varepsilon_{r1} = 2.17$, $\varepsilon_{r2} = 6.12$, loss tangent of substrate 1 = 0.0009, loss tangent of substrate 2 = 0.00022, $h_1 = 1.5748$ mm, $h_2 = 1.27$ mm.

The antenna is designed to operate at a frequency of 2.7 GHz. The range of frequencies around the operating frequency in which the antenna can operate depends on the frequencies at which the VSWR value is less than 2. The design objective is to maximize its VSWR BW and get the resonance at desired 2.7 GHz.

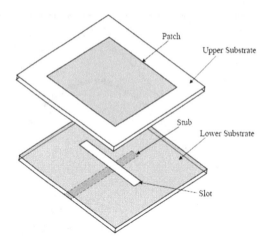

Fig. 5.14 Geometry of the aperture coupled microstrip antenna [97]

Experiment and Results

In this example, we used PSO algorithm in order to obtain optimal design parameters of the microstrip antenna. The hybrid approach is named as Support Vector driven Particle Swarm Optimization (SVPSO). The fitness function used for optimization is:

$$f = \alpha(|f - CF|^{c1}) - \beta(BW^{c2}) \tag{5.10}$$

$$\alpha + \beta = 1 \tag{5.11}$$

where CF is the frequency at which resonance is obtained and BW is the VSWR bandwidth measured at level 2 from VSWR vs. frequency plot. The parameters $c1$, $c2$, α and β were chosen as per the requirement of whether a large bandwidth is desired or proper matching is desired. The variables $c1$ and $c2$ allow us to model a non-linear correspondence between $|f - CF|$ and BW. The function f attains the minimum value when frequency is matched, and the bandwidth is high. The design parameters corresponding to the minimum value of f would thus be the optimal design parameters.

The accuracy of SVM models of the microstrip antenna is compared with FF-NN model generated using same dataset it is shown in Table 5.5. The optimization was carried out using presented SVPSO approach and the results are shown in 5.6. The optimized result is compared with the best result available in the training data set.

Table 5.5 Accuracy of modeling for aperture coupled MSA [97]

Models	MSE	ARE
Feedforward ANN	0.00309	0.018
SVM	2.025e-5	0.001663

Table 5.6 Results of Optimization for aperture coupled MSA [97]

	Bandwidth	VSWR	Operating Frequency
Without optimization*	0.03 GHz	1.20	2.6 GHz
After optimization	0.05 GHz	1.22	2.7 GHz

*Best Bandwidth, VSWR pair available in training set.

5.6 Concluding Remarks

In this chapter, an effective modeling of microwave components using SVM framework is presented and used to model a one port microstrip via, an aperture coupled microstrip antenna, and a circular polarized microstrip antenna. The models obtained with SVM are compared with conventional ANN models prepared for the same components and are found to be more accurate. In this experiment, grid regression was used to select parameters of SVM. However, nonconventional optimization algorithms such as GA, and PSO can also be used to obtain best combination of these parameters.

We have presented a hybrid approach combining SVR model with GA and PSO. This approach with different optimization algorithms (GA and PSO) is demonstrated for the design of two microstrip antennas respectively for each optimizer. The design of circular polarized microstrip antenna was carried out using SVGA approach, while the design of aperture coupled MSA was carried out using SVPSO approach. The advantage in applying this hybrid approach is that number of EM simulations required can be restricted to the number of experimental data required to generate SVR models. This number is very less compared to the number of simulations required, if EM simulator is invoked in the optimization loop of GA/PSO. The main requirement in performing this is to obtain desired accuracy of the model. As we observed from the experiments, SVR gives better accuracy compared to ANN models.

Though we have considered the design of microstrip antennas as a case study in the present work, the hybrid method presented here can well be applied to the design of many other microwave components. It is highly advantageous when an approximate model with the desired accuracy can be obtained with less computational expense.

Design of a Nonlinear Taper Using SI Based Algorithms

6.1 Overview

Tapered transmission lines (tapers) are common means of transmitting power from one part/device to another part/device in many microwave and millimeter-wave devices. A taper are required to transform the output of a standard waveguide to oversized waveguide components. The design of taper should be carried out in such a way that the characteristic impedances at both the ends match. Two basic types of cross-section tapers are straight taper and variable (nonlinear) taper. In the straight taper, the taper angle is fixed throughout the length and abrupt discontinuities occur at both the ends, while in variable taper the taper angle is smoothly varied along the length of the taper. The advantage of a nonlinear taper is that the conversion of power to unwanted (spurious) modes is very less compared to straight taper [110]. In this chapter, the design and optimization of a nonlinear taper for use in a specific high power gyrotron is presented.

Gyrotron (a type of millimeter-wave source used in fusion reactors) is capable of providing hundreds of kilowatts of power at microwave and millimetric wavelengths [110, 111]. The output power generated using a gyrotron is very high, ranging from long pulse to Continuous Wave (CW). High power gyrotrons are mainly used for plasma heating in thermonuclear fusion reactors, Tokamaks and stellarators. In addition, they are used in a variety of ISM applications ranging from spectroscopy, material processing to plasma diagnostics. The output system of a gyrotron consists of an interaction cavity, output taper, a quasi-optical mode converter, and an RF window [110, 112]. In gyrotrons, the nonlinear taper is used to connect the interaction region with the output waveguide system. The main challenge in the design of a nonlinear taper for gyrotron is that it requires very high transmission (above 99%) with very less spurious mode generation. Due to high output power, even 1% of reflections can cause severe damage to the entire system. For this reason, although it is a simple component, it has to be designed carefully. This is a typical design application where the accuracy requirement is very

N. Chauhan, M. Kartikeyan, and A. Mittal: Soft Computing Methods, SCI 392, pp. 69–82.
springerlink.com © Springer-Verlag Berlin Heidelberg 2012

high. It is necessary that a proper optimization tool be chosen in order to find a global optimum from the complex and nonlinear design space.

Evolutionary Algorithms (EAs) such as GA, and SI based algorithm such as PSO have proved to be effective and promising tools for search and optimization in multidimensional feature space. The applicability of PSO for gyrotron mode convertor application is illustrated in [116]. In this chapter, we investigate the role of a more recent SI based technique, bacterial foraging optimization, for the design of a nonlinear taper. Since, the original bacterial foraging optimization has several drawbacks (as mentioned in next section), an effort has been made to generate a modified BFO algorithm by introducing some concepts of PSO algorithm in it. In this chapter, modified BFO has been compared with the standard PSO algorithm on numerical benchmarks first, and later on both are used for the design of a nonlinear taper as a case study application of high power microwave and millimeter wave devices [113–115, 117].

Review of Conventional Approaches for Taper Design

Various nonlinear tapers used for matching purposes are triangular taper, exponential taper and Chebyshev taper [118, 119]. Few of tapers that are generally employed in microwave systems have been designed by Klopfenstein [120], and Hecken [121], etc. Flügel and Kühn developed programs for computer-aided analysis and design of Dolph-Chebyshev and modified Dolph-Chebyshev tapers [122] for use in gyrotrons. The theoretical evaluation of two nonlinear tapers of raised-cosine type was presented by Lawson in [123]. He showed that under some conditions, raised-cosine profile yields less mode conversion than modified Dolph-Chebyshev profile. In the present work, we have considered the design and optimization of a raised-cosine taper for use in high power gyrotron.

This chapter is organized as follows. Section 6.2 discusses drawbacks of standard BFO algorithm, its modification for improved convergence, and its comparison with standard PSO algorithm using benchmark test functions. Section 6.3 presents the use of both algorithms for the design and optimization of a specific nonlinear taper. Finally, section 6.4 presents concluding remarks based on the performance of both SI based algorithms and their role in the design of high power millimeter-wave devices.

6.2 Modified Bacterial Foraging Optimization

The standard bacterial foraging optimization algorithm as discussed in subsection 2.3.2 when tested on multidimensional benchmark functions is observed to show a poor convergence compared to standard PSO [79]. One of the reasons (as compared to standard PSO) is that classical BFO ignores the effects of global swarming. Moreover, all the bacteria are assumed to have same swim length in their chemotaxis process. However, the use of variable swim length according to their relative distance from the bacterium with highest nutrient position may improve convergence.

In order to improve the convergence, a modified BFO (MBFO) algorithm is presented by considering few changes in the standard BFO. The modifications applied to the standard BFO algorithm are as follows [117]:

- In classical BFO, it is assumed that bacteria will move in a random direction in every chemotaxis loop. But, in nature, a bacterium can remember the nutrient concentration in its previous postion [34]. Based on this knowledge, a bacterium compares its current nutrient concentration with that of the previous concentration. This information is stored in a memory array which is a $S \times Dim$ dimensional vector. In the presented modification, each time a bacterium encounters a favorable environment, it remembers the direction in which it moved, so that in the next swim step it can move in the same direction as it is more probable to encounter a favorable environment. If at some point it reaches a toxic place (higher value of J_i) it moves back to its previous position and from this point it generates a random direction (i.e., it tumbles) and moves in a new direction. In this way the bacteria can reach highest nutrient concentration (minimum J_i) quickly.
- In the second modification to the standard BFO algorithm, after undergoing a chemotactic step, the position of bacteria is modified by applying PSO operator as suggested in [79]. In this phase, the bacterium is stochastically attracted towards the global best position found so far in the entire swarm. The 'social' component of PSO is only used, ignoring 'cognitive' component, as the local search in different regions of the search space is already taken care of by the chemotactic steps. The velocity and position update equations used in applying PSO operator are as follows:

$$V^i_{new} = wV^i_{old} + C_1 r_1 (x^{gbest} - \theta^i(j, k, l)) \tag{6.1}$$

$$\theta^i_{new}(j+1, k, l) = \theta^i_{old}(j+1, k, l) + V^{id}_{new} * Vary \tag{6.2}$$

where v_{id} is the velocity of ith bacterium in dth dimension, w, C_1, $Vary$ are constants, r_1 is a uniformly generated random number.
- The chemotactic step-size is varied in accordance with fitness [124] as:

$$A = \frac{K}{|\, Jbest - J(i, j+1, k)\,|} \tag{6.3}$$

$$C(i) = \frac{1}{1+A} \tag{6.4}$$

where K is constant. For bacteria far away from best, $C(i)$ should be near to 1, so it results in a large step size. For bacteria close to J_{best}, $C(i)$ should be near to 0, so that they can converge to global minima without much oscillations. In the process, $K{=}400$ was observed to give satisfactory results.

The detailed flowchart of the MBFO algorithm is shown in Fig. 6.1, whereas the detailed algorithm is stated as below.

ALGORITHM: *The Modified Bacterial Foraging Optimization Algorithm*
Initialization:

1. Initialize parameters Dim, S, N_c, N_s, N_{re}, P_{ed}, N_{ed}, $C(i)$ with ($i = 1, 2, ..., S$), θ^i, w, and C_2 where,
   Dim: Dimension of the search space,
   S: Number of bacteria in the population,
   N_c: Number of chemotactic steps,
   N_{re}: Number of reproduction steps,
   N_s: Length of swimming,
   N_{ed}: Number of elimination-dispersal events,
   P_{ed}: Probability of elimination-dispersal events,
   $C(i)$: Size of the step taken in the random direction specified by the tumble,
   $\theta^i(j, k, l)$: Position vector of the i-th bacterium, in j-th chemotactic step, k-th reproduction step and l-th elimination-dispersal event,
   w : **Inertia weight,**
   C_2 : **Social coefficient of PSO algorithm.**

Iterative loops:
2. Elimination-dispersal loop: $l{=}l{+}1$
3. Reproduction loop: $k{=}k{+}1$
4. Chemotaxis loop: $j{=}j{+}1$

   a. For $i = 1, 2, ..., S$, take a chemotactic step for bacterium i as follows:
   b. Compute fitness function $J(i, j, k, l)$, and then let,
      $$J(i, j, k, l) = J(i, j, k, l) + J_{CC}(\theta^i(j, k, l), P(j, k, l)).$$
   c. Let $J_{last} = J(i, j, k, l)$ to save this value since we may find a better cost via a run.
   d. *Tumble:* Generate a random unit vector $\phi(i)$ with each element $m(i)$, $m{=}1, 2,..., Dim$, a random number on $[-1, 1]$.
   e. *Move:* Following the eq. (2.6).
   f. Compute $J(i, j{+}1, k, l)$ as, $J(i, j{+}1, k, l) = J(i, j{+}1, k, l) + J_{CC}(\phi^i(j{+}1, k, l), P(j{+}1, k, l))$.
   g. *Swim:* Consider only the i-th bacterium is swimming while the others are not moving, then,
      i. Let $m{=}0$ (counter for swim length)
      ii. While $m < N_s$ (if have not climbed down too long)
         • Let $m = m + 1$
         • If $J(i, j + 1, k, l) < J_{last}$ (if doing better) then,
           $J_{last} = J(i, j+1, k, l)$ and $\theta^i(j+1, k, l) = \theta^i(j+1, k, l) + C(i)\phi(i)$

and use this $\phi^i(j+1,k,l)$ to compute the new $J(i,j+1,k,l)$ as in (f).

- **Else (if position is not better)**
  Tumble to generate new unit random vector $\phi(i)$
  $J_{last} = J(i,j+1,k,l)$ **and** $\phi^i(j+1,k,l) = \phi^i(j,k,l) + C(i)\phi(i)$
  End if
- **Modify next step size $C(i)$ as,**
  $C(i) = 1/(1+A)$, **where** $A = K/|J_{best} - J(i,j+1,k,l)|$

iii. End while loop

h. If $i \neq S$ then $i = i+1$ and go to (b) to process $(i+1)$ bacterium.

5. If $j < N_c$ go to step 4 (i.e., continue chemotaxis as the life of the bacteria is not over).

6. **Modify positions of bacteria by applying PSO operator as,**

   a. **For $I = 1,2,...,S$**
      i. **Update θ_{gbest} and $J_{best}(i,j,k,l)$**
      ii. **Update velocity and position of i-th bacterium as,**
      $V_i^{new} = w \cdot V_i^{old} + C_2 r_1 (\theta_{gbest} - \theta^i(j,k,l))$
      $\theta^i(j,k,l) = \theta^i(j,k,l) + V_i^{new}$
   b. **End for**

7. *Reproduction:* Sort bacteria in ascending of their fitness value (J). Now, let $S_r = S/2$. The S_r bacteria with highest cost function (or fitness) values (J) die and the other half of bacteria population with the best values split (and the copies that are made are placed at the same location as their parent).

8. If $k < N_{re}$, go to step 3. We have not reached the specified number of reproduction steps. So we start the next generation of the chemotaxis loop.

9. *Elimination-dispersal:* For $i = 1,2,...,S$, eliminate and disperse each bacterium with probability P_{ed}. (If any bacterium is eliminated, then disperse other bacterium to random location in optimization domain in order to keep the number of bacteria in population constant.) If $l < N_{ed}$ then, go to step 2; otherwise end.

Parameter Selection. The ability of BFO in exploring global optima is greatly dependent on the choice of parameters such as $C(i), N_c, N_{re}, N_{ed}, N_s$. In classical BFO, Passino took $C(i) = 0.1, i = 1,..,S$. The values of parameters selected in our experiments for MBFO algorithm on five benchmark functions are shown in Table 6.1.

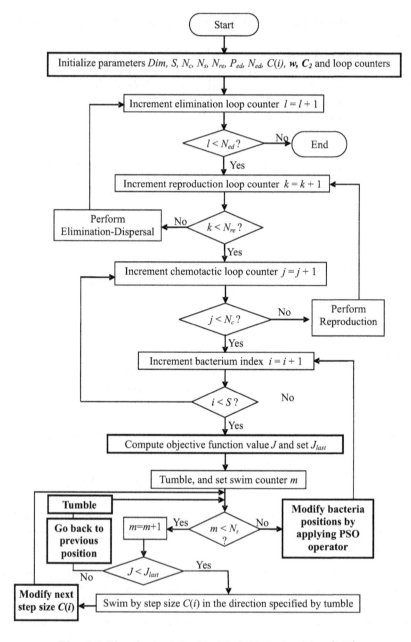

Fig. 6.1 Flowchart of the Modified BFO algorithm [117]

Table 6.1 Values of user defined parameters of BFO [23]

Name of parameter	Value of parameter
N_c	40
N_{re}	5
N_{ed}	5
P_{ed}	0.25
N_s	20
$C(i)$	variable
$Vary$	0.075 for range of search is small (say<100), 1 otherwise

Table 6.2 Benchmark functions [81] (© 2005 IEEE)

Function name	Function
De Jongs's sphere function	$f_1 = \sum_{i=1}^{n} x_i^2$
Rosenbrock function	$f_2 = \sum_{i=1}^{n-1} [100(x_{i+1} - x_i^2)^2 + (x_i - 1)^2]$
Generalized Rastrigin function	$f_3 = \sum_{i=1}^{n} [x_i^2 - 10\cos(2\pi x_i) + 10]$
Generalized Griewank function	$f_4 = \frac{1}{4000} \sum_{i=1}^{n} x_i^2 - \prod_{i=1}^{n} \cos\left(\frac{x_i}{\sqrt{i}}\right) + 1$
Schaffer's f6 function	$f_5 = 0.5 + \frac{(\sin\sqrt{x_1^2 + x_2^2})^2 - 0.5}{(0.1 + 0.001(x_1^2 + x_2^2))^2}$

Table 6.3 Initialization range, search range, and error tolerance [81] (© 2009 IEEE)

Test function	Range of search	Range of initializations	Error tolerance
f_1	$[-100, 100]^n$	$[50, 100]^n$	0.01
f_2	$[-100, 100]^n$	$[50, 100]^n$	0.01
f_3	$[-10, 10]^n$	$[2.56, 5.12]^n$	0.01
f_4	$[-600, 600]^n$	$[300, 600]^n$	0.01
f_5	$[-100, 100]^n$	$[15, 30]^n$	0.00001

6.2.1 Experiment on Benchmark Functions

Experimental Settings

The performance of the MBFO algorithm was evaluated on a test bed of same five benchmark functions that were used in Chapter 4 and are reproduced here in Table 6.2. The search range for these functions and the range of initializing bacteria positions are similarly reproduced and shown in Table 6.3 for convenience. Remember that asymmetric initializations are used for these functions in order to consider the practical situations. All these functions were tested with 10, 20 and 30 dimensions, except the last Schaffer f6 function which is two dimensional. For De Jong's Sphere function, the maximum number of iterations considered to stop the BFO algorithm were 1000, 2000 and 3000 for dimension 10, 20 and 30 respectively. While, they were considered to be 3000, 4000 and 5000 for 10, 20 and 30 dimensions respecively for Generalized Rastrigrin, Generalized Rosenbrock and Generalized Griewank

Table 6.4 Performance of MBFO on benchmark functions and its comparison with standard PSO algorithm [117]

Fn.	Dim.	Average achieved optimum value (Standard deviation)		Number of successes (Average generation of success)	
		PSO	MBFO	PSO	MBFO
f_1	10	0.01(−)	0.14(1.28)	50(566)	38(374)
	20	0.01(−)	0.01(−)	50(1291)	50(1274)
	30	0.01(−)	0.015(0.00)	50(2081)	49(2338)
f_2	10	25.19(64.34)	0.14(1.28)	1(2365)	3(1955)
	20	131.94(269.92)	2.46(6.62)	0(−)	0(−)
	30	186.85(370.73)	3.88(2.41)	0(−)	0(−)
f_3	10	2.30(1.55)	59.35(41.44)	4(1904)	0(−)
	20	14.20(4.68)	58.37(24.73)	0(−)	0(−)
	30	31.44(7.15)	82.26(26.49)	0(−)	0(−)
f_4	10	0.075(0.035)	0.01(0.00)	0(−)	50(875)
	20	0.026(0.026)	0.01(0.00)	22(2339)	50(2349)
	30	0.02(0.015)	0.01(0.00)	27(3222)	50(3790)
f_5	2	0.00097(0.0029)	0.27(0.27)	45(532)	0(−)
Total successes (Total iterations)				249(14300)	290(12955)

functions. The maximum iterations considered for Schaffer f6 function was 1000. The maximum error tolerance considered for all functions was 0.01 except for Schaffer f6 function in which it was considered to be 0.00001. The MBFO algorithm was stopped when either the error criterion or maximum iterations were reached.

Simulation Results on Benchmark Functions

The results of testing MBFO on test functions are shown in Table 6.4. Due to probabilistic nature of evolutionary algorithms, the results shown here are an average of 50 runs of the algorithm for each test case. In order to compare the performance of MBFO algorithm, similar results were obtained using standard PSO algorithm. Here, three metrics are used for comparative study, (i) quality of solution (measured in terms of average achieved optimum value and its standard deviation), (ii) convergence speed (which is measured in terms of number of successful runs out of 50 runs), and (iii) average number of iterations for only successful runs.

The comparative results of MBFO and PSO on test functions show that MBFO outperforms PSO on Rosenbrock's function and Generalized Griewank function, while the standard PSO performed well on Generalized Rastrigrin and Schaffer f6 function. The total number of successes obtained with MBFO is higher than those obtained with standard PSO algorithm.

6.3 Design of a Nonlinear Taper for High Power Gyrotrons

6.3.1 Design Considerations

In this work, the design of a nonlinear taper is carried out for a 42 GHz, 200 kW, CW gyrotron operating in the $TE_{0,3}$ mode with axial output collection. This work was carried out as a part of a specific gyrotron development project undertaken currently in India. A schematic diagram of a raised-cosine taper considered in this work is shown in Fig.6.2. The analysis of the taper was carried out using a dedicated scattering matrix code [125], as it was found to be fast and accurate for taper analysis. The tapered parts were divided like a flight of stairs (see Fig. 6.2). The scattering coefficient of each step was calculated by using a dedicated scattering matrix code. The synthesis of the raised-cosine taper profile was carried out using the following formulae [126]:

$$\alpha = -1.0 + 2.0 \left(\frac{i}{l}\right)^{\gamma} \tag{6.5}$$

$$r(z) = \frac{r_2 - r_1}{2} \cdot \left(\alpha + \frac{1}{\pi} sin(\pi\alpha)\right) + \frac{r_2 - r_1}{2} \tag{6.6}$$

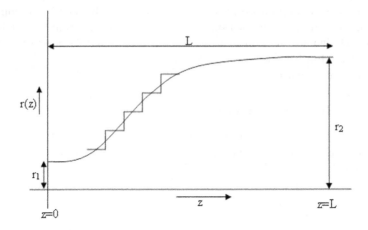

Fig. 6.2 The raised-cosine taper profile [122, 110].

$$r = r_1 + r(z) \tag{6.7}$$

The goal here is to find optimum shape for which maximum transmission occurs with minimum spurious mode contents. For a nonlinear taper to be used in gyrotron, its power reflection and spurious mode contents should be less than 1 %. The optimized performance of the taper was obtained using both standard PSO (as described in subsection 2.3.1) algorithm and MBFO algorithm as presented in section 6.2. In this design optimization task, it was identified that a geometrical parameter (γ), in Eq. (6.5), is very important for determining the shape of the raised-cosine taper. The effect of varying γ parameter on the transmission characteristics of the nonlinear taper is shown later in this section. In this optimization process, parameter γ along with three other nominal parameters namely length of the taper (L), radius of the taper at output end (r_2), and the number of straight sub-sections (N) were considered as design parameters. In the experiment, the radius of the taper at input end (r_1) was fixed at a constant value of 13.991 mm. The range of design parameters considered for the design are mentioned in Table 6.5. The goal during the optimization was to obtain maximum transmission coefficient (i.e., S_{21}-parameter), operating in $TE_{0,3}$ mode with minimized spurious mode content. The fitness function considered for minimization in this problem is as follows:

$$f_{obj} = -S_{21(TE_{0,3})} \tag{6.8}$$

where S_{21} is transmission coefficient for $TE_{0,3}$ mode. The minimization of spurious mode contents are also obtained by maximizing transmission coefficient, since the total power remains constant.

Table 6.5 Range of design parameters [113]

Design Parameter	Range (in mm)
Length (L)	200-350
Radius at output end (r_2)	35-45
Number of sections (N)	50-500
Gamma (γ)	0.1-1.0

Table 6.6 Optimized values of taper parameters [117, 113]

Name of Design Parameter	Optimized Value	
	PSO	MBFO
Length (L)	349.99 mm	315.45
Radius at output end (r_2)	37.32 mm	35.0
Number of sections (N)	208	469
Gamma (γ)	0.504	0.517
Transmission	99.87 %	99.85 %

6.3.2 Optimized Results Using PSO and MBFO Algorithms

The optimized design parameters and corresponding transmission coefficient using PSO and MBFO are shown in Table 6.6. It can be observed that excellent transmission of 99.87% and 99.85 % are obtained using both the algorithms respectively. The algorithm was executed with a swarm size of 10 particles and 20 bacteria respectively, for about 100 iterations using both algorithms. The iterations considered for the design were found sufficient for the convergence of the swarm.

During the experiment, the effects of varying individual parameters such as gamma parameter (γ), radius at the output end of the taper (r_2), and length of the taper (L) on the taper synthesis and the transmission coefficient were also observed. The effect of the gamma parameter is shown in Fig. 6.3, while the effect of radius at output end of the taper on the transmission coefficient is shown in Fig.6.4. It can be observed that the value of parameter gamma (γ) plays significant role on the transmission efficiency of the taper. It was observed during the experiments that the value γ between 0.5 and 0.7 leads

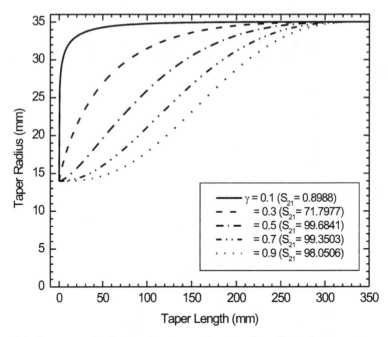

Fig. 6.3 Contours of raised-cosine taper showing the effect of parameter gamma (γ)(L=350 mm, r_2 =35.0 mm, N=466) [113]

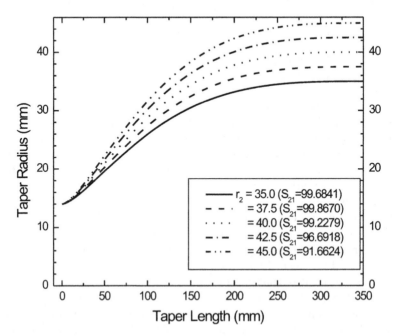

Fig. 6.4 Contours of raised-cosine taper showing the effect of output radius (r_2) ($L = 350$ mm, $\gamma = 0.5$, $N = 466$) [113]

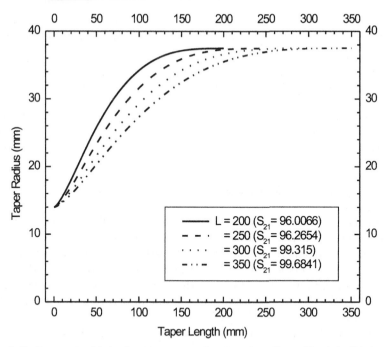

Fig. 6.5 Contours of raised-cosine taper showing the effect of length (L) ($r_2 = 35$ mm, $\gamma = 0.5$, $N = 466$) [113]

Table 6.7 Percentage of transmitted and reflected power for various modes (Frequency: 42.0 GHz) [23]

Mode	Reflection[%]	Transmission[%]
$TE_{0,1}$	0.00001	0.01673
$TE_{0,2}$	0.00001	0.00012
$TE_{0,3}$	0.00044	99.86609
$TE_{0,4}$	0.01239	0.00762
$TE_{0,5}$	0.00123	0.00028
$TE_{0,6}$	0.00036	0.00010
$TE_{0,7}$	0.00015	0.00019
$TE_{0,8}$	0.00007	0.00015
$TE_{0,9}$	0.00004	0.00005
$TE_{0,10}$	0.00002	0.00000

to transmission above 99%. It can also be observed that output radius of less than 40 mm leads to the desired transmission range and gives optimum transmission at 37.5 mm.

The percentage of transmission for various taper lengths within the desired range are shown in Fig. 6.5. It can be observed that as we increase the taper length, the transmission also increases. In addition, the percentage of transmission and reflection of the main mode $TE_{0,3}$ along with its azimuthal neighbours for an optimized run (with r_1=13.991, r_2=37.32, L=349.99, γ=0.504, N=208) are shown in Table 6.7.

6.4 Concluding Remarks

The design and optimization of a raised-cosine type nonlinear taper profile for a 42 GHz, 200 kW, CW gyrotron was carried out using two swarm intelligence based algorithms namely standard PSO algorithm and a modified version of BFO algorithm. A dedicated scattering matrix code [125] was used for the fast and accurate analysis of taper during design process. The selection of PSO, and MBFO parameters were carried out in accordance with previous experimental investigations.

The optimized results show excellent matching obtained for a raised-cosine taper which confirms effectiveness of the presented methods for the design of nonlinear tapers. The time required for the optimization was 20 and 40 minutes (approximately) respectively for PSO and MBFO algorithms on a latest workstation. It can be concluded here that both the algorithms can find global optimum from the nonlinear design space of a complex taper design problem, which requires perfect design with very high transmission (above 99%). Excellent transmission of 99.87% and 99.85% were obtained with both PSO and MBFO algorithms in very small number of iterations. This shows the power of swarm intelligence based algorithms in solving critical high power microwave design problems. Finally, a parametric analysis of the taper was carried out by varying individual design parameters in a discrete fashion while keeping other parameters constant.

Multiobjective Optimizations for the Design of RF Windows

7.1 Overview

High power microwave and millimeter-wave sources such as gyrotrons, klystrons, and other gyro-devices produce huge amount of output power at wavelengths from microwave to millimeter-wave range [110, 127]. Among many components, RF-window is an important component of the output system of these devices. It serves as a barrier between the vacuum side of the device and the output transmission line. One of the important tasks in the design of RF-window is the selection of suitable window material with required characteristics/features [128]. The material selected must have high power handling capability. It should withstand thermal and mechanical stresses, should avoid flashing/arching and puncturing/physical damage. So, care must be taken in selecting proper window material with low loss tangent, high thermal conductivity, suitable mechanical strength, and perfect design to minimize power reflections and absorption for better transmission [110]. An important challenge is that a perfect design should be obtained that minimizes power reflections at desired frequency. In this chapter, we present the design of two types of RF windows - double disc window and pillbox-type window. Double disc windows are used in gyrotrons and other gyro-devices [110, 129], while RF-window of pillbox-type is usually employed in conventional microwave sources, such as, klystrons [130].

7.1.1 Review of Conventional Designs of RF-Window

A few designs reported in the literature for the design and analysis of single disc, double disc, and pillbox-type RF-windows used in high power microwave and millimeter-wave sources are as follows. Thumm describes various types of high power windows and summarizes the early development status of high-power millimeter-wave windows with emphasize on the use of CVD diamond [128]. Yang, *et al.* [129] have presented analysis of conventional single disc

N. Chauhan, M. Kartikeyan, and A. Mittal: Soft Computing Methods, SCI 392, pp. 83–95.
springerlink.com   © Springer-Verlag Berlin Heidelberg 2012

window, frequency tunable double disc window, and ultra-broadband Brewster window. They optimized geometrical parameters of these windows to obtain suitable transmission characteristics. Yang, *et al.* [131] also investigated the influence of some window parameters such as mechanical tolerance of the disc thickness, variation of distance between two discs, and frequency shift during gyrotron pulse on the transmission characteristics of millimeter waves. They show that the power reflections is a critical issue; and an accurate tuning is required for the design of double disc window to keep reflections below 1%. Jöstingmeier, *et al.* [130] present a systematic method for designing 75 MW S-band pillbox-type RF-window. They used an additional inductive iris to make use of ceramic disc of arbitrary thickness. In their work, the bandwidth was optimized by varying the thickness of the disc.

The major two objectives in the design of RF-window are to match the working frequency at a given target, and to obtain minimum reflections at that frequency. Sometimes during the design, while minimizing the reflection, frequency gets shifted from the target frequency. This is an example where optimizing one objective leads diverse effect on the other objective. Soft computing optimization techniques can handle design with multiple objectives which may be conflicting in nature. In this chapter, we demonstrate the use of Multi-Objective Particle Swarm Optimization (MOPSO) for optimizing design parameters of disc-type RF-windows [132, 133]. The MOPSO is selected to treat minimization of power reflection around resonant frequency (this also maximize BW around it) and exact matching at desired frequency as separate objectives. This approach gives a set of solutions, which optimally balances the trade-off between objectives. The designer gets chance to select a design from available set of solutions according to the requirement.

The organization of the chapter is as follows. Section 7.2 describes the concept of MOPSO and the use of MOPSO in microwave design so far. Section 7.3 presents the design of three RF-windows for specific high power microwave sources using MOPSO methodology. Finally, section 7.4 presents the concluding remarks.

7.2 Multi-objective Particle Swarm Optimization

7.2.1 *Multi-objective Optimizations*

It is observed that many real-time problems have more than one objective. For these problems, it is desired to find a solution that optimally balances the trade-off between multiple objectives. The goal of the Multi-Objective Optimization (MOO) methods is to obtain a set of solutions, called *Pareto-optimal set*, which optimally balance the tradeoff between multiple objectives based on the concept of *Pareto dominance* [29, 134]. A solution X is called to dominate solution Y, if $f(X) \leq f(Y)$ for all individual objectives and $f(X) < f(Y)$ for at least one objective. A set of objective vectors in objective space corresponding to Pareto-optimal set is referred to as *Pareto front*. Pareto front can be depicted as a graph in which each objective is mapped along a dimension, and the points in the graph represent optimum solutions.

The concept of MOO with evolutionary algorithms such as GA has been explained in detail in [141]. Similar to genetic algorithms, MOO is also implemented with particle swarm optimization by several researchers. A comprehensive summary of most MOPSO methods is given in [29].

7.2.2 Review of Microwave Design Using MOPSO

Microwave researchers have used various MOPSO implementations for the design of various electromagnetic and antenna design problems. For instance, Gies and Rahmat-Samii used Vector-Evaluated PSO (VEPSO) for the design of antenna array [135]. They applied VEPSO to optimize beam efficiency and Half Power Beam Width (HPBW) of a 16-element radiometer array antenna. A drawback with this algorithm is that handling of more than two objectives is not possible. Later, Xu and Rahmat-Samii also used adaptive MOPSO by applying it to two antenna design problems: synthesis of 16-element antenna array which is optimized for tradeoff between beam efficiency and HPBW, and optimization of shape reflector antenna for high gains of multiple feeds [50].

In this chapter, a specific implementation of MOPSO developed by Raquel and Naval [136] has been used for the design of three different RF-windows. This implementation is based on a mechanism called *crowding distance*. Crowding distance is a metric which provides estimate of density of solutions surrounding a particular solution. This mechanism is incorporated in PSO algorithm for the selection of global best and for the deletion of solutions from external archive of non-dominated solutions. This method is suggested to be highly competitive in converging towards the Pareto-front. It is also found to give well distributed non-dominated solutions on optimization test problems. In the presented work, this method has been used for finding optimum Pareto front for the design of double disc RF-windows and pillbox-type RF-window for use in high power microwave and millimeter wave devices such as gyrotrons and klystrons respectively. The algorithm is aimed to find optimum physical dimensions for both the types of RF-windows. This MOPSO with crowding distance has been used to achieve optimized trade-off between matching of desired resonant frequency and minimizing the power reflections which, in this case, is obtained by maximizing the bandwidth around resonant frequency.

7.3 Design of Disk-Type RF Windows Using MOPSO

In this section, three design problems of disc-type RF-windows namely a double disc window for 42 GHz and 170 GHz gyrotrons, and a pillbox-type window for 2.856 GHz klystron have been considered [132]. Out of the above mentioned windows, the design of a RF-window for a 42 GHz gyrotron is a part of the gyrotron development project currently undertaken in India. The detailed design specifications for all three windows, their desired working frequency, window materials and their dielectric properties used for each type of window are shown in Table 7.1.

In all three experiments, an MOPSO with crowding distance [136] was used for getting optimum Pareto front. The MOPSO algorithm was executed with a swarm size of 20 for 500 iterations in each case. The size of the archive for storing optimum Pareto sets during the execution was kept at 20 in each case. The values of c_1 and c_2 parameters of MOPSO were considered to be 2 in this implemetation. The value of inertia weight w was varied from an initial value of 0.9 to a final value of 0.4 during the iterations. The boundary constraints for the design parameters are handled by imposing boundary cutoff limits as described in Chapter 4.

Table 7.1 Specifications considered for the design of RF-windows [132] (© 2009 IEEE)

Type of Window	Frequency [GHz]	Window Material	Dielectric Constant & Loss Tangent
Double disc	42.0	Sapphire	$^*\epsilon_r$=9.41, $tan\delta$=5.5e-05
		SiN	$^*\epsilon_r$=7.9, $tan\delta$=1.0e-04
Double disc	170.0	CVD-Diamond	$^*\epsilon_r$=5.67, $tan\delta$=4.0e-05
Pillbox	2.856	AL995	$^*\epsilon_r$=9.37, $tan\delta$=9.0e-05

* Note: Dielectric properties for Sapphire and SiN are given for T=300K, f =42 GHz from Kartikeyan et al. [137] and for CVD-Diamond is given for T=300K, f=170 GHz from Thumm [128], and for AL005 is given for T=300K, f =2.856 GHz from Singh, et al. [138].

7.3.1 Design of a Double Disc RF-Windows for a 42 GHz Gyrotron

In this design example, a double disc RF-window is considered for a 42 GHz, 200 kW, CW gyrotron to work in the $TE_{0,3}$ mode. The schematic diagram of the double disc window considered for the design is shown in Fig. 7.1. The design parameters considered for optimization are disc thickness ($D_1 = D_2$), distance between two discs (t_{dd}), and the disc radius (r_{dd}). The same disc thickness for both the discs are considered in the experiment. The range of parameters considered for the design optimization using MOPSO are shown

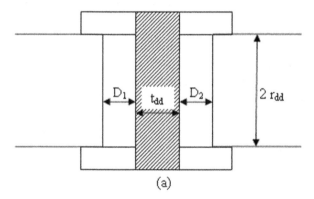

(a)

Fig. 7.1 Schematic diagram of double disc RF-window [133, 132] (© 2009 IEEE)

in Table 7.2. In this design experiment, the initial guess for the value (and also to get the range) of disc thickness for the TE modes is obtained from Eq. 7.1 [129].

$$t_{dd} = \frac{\pi}{\sqrt{(\frac{2\pi f \sqrt{\epsilon_r}}{c})^2 - (\frac{\chi_{mn}}{R_{win}})^2}} \tag{7.1}$$

Here, f is desired resonant frequency, R_{win} specifies radius of the disc, and χ_{mn} is the Bessel root for the $TE_{m,n}$ mode.

Table 7.2 Range of design parameters of a double disc window for 42 GHz gyrotron [132, 133] (© 2009 IEEE)

Design parameters	Range [mm]
Disc thickness ($D_1 = D_2$)	2.0-3.0
Distance between two discs (t_{dd})	2.0-4.0
Radius (r_{dd})	35.0-45.0

The two objective functions considered for the design using MOPSO are as follows:

$$f_1 = |f - f_{desired}| \tag{7.2}$$

$$f_2 = \frac{1.0}{BW + 1.0} \tag{7.3}$$

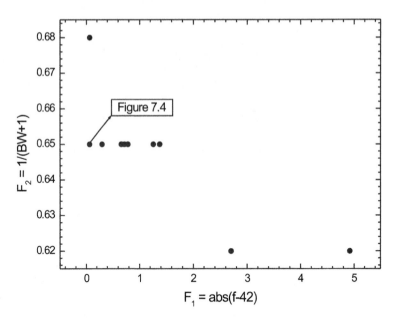

Fig. 7.2 Pareto front of a double disc window with Sapphire disc material for a 42 GHz gyrotron [132] (© 2009 IEEE)

where f is the resonant frequency (in GHz) obtained in each execution, $f_{desired}$ is the desired resonant frequency for the window (in GHz) and BW specifies the bandwidth measured at -20 dB from the reflection response curve (S_{11}-parameter) around resonant frequency. Here, the function f_1 tries to match the frequency response at desired frequency, while the function f_2 minimizes the reflections around resonant frequency by maximizing the bandwidth.

In this design, two different experiments have been carried out considering two different disc materials, namely, Sapphire and SiN. The dielectric constant and loss tangent for these materials are given in Table 7.1. During the design process, the analysis of the double disc window was carried out following a dedicated code developed by [110, 139]. The code was invoked in the optimization loop of MOPSO algorithm as an objective function in order to get the analysis response each time.

The Pareto fronts showing the trade-off between two objectives for the double disc window using disc materials Sapphire and SiN are shown in Fig. 7.2 and Fig. 7.3 respectively. Each symbol in the graph specifies different optimum designs with respect to objective functions in the objective space. The transmission and reflection responses for the best designs obtained in each case (for Sapphire and SiN materials) are shown in Fig. 7.4 and Fig. 7.5 respectively. The responses using the same optimum design parameters were also obtained with a scattering matrix code Cascade 3.0 [140] for the purpose of comparison. These responses are also shown in the same figures.

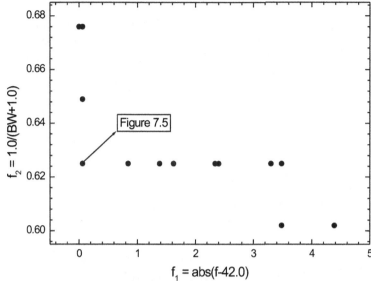

Fig. 7.3 Pareto front of a double disc window with SiN disc material for a 42 GHz gyrotron [132] (© 2009 IEEE)

Table 7.3 Range of design parameters of a double disc window for 170 GHz gyrotron [132] (© 2009 IEEE)

Design Parameters	Range [mm]
Disc thickness $(D_1=D_2)$	1.0-2.0
Distance between two discs (t_{dd})	3.0-4.0
Radius (r_{dd})	45.0-55.0

The optimized values of design parameters in each case are specified along with their responses (see caption of Fig. 7.4 and Fig. 7.5). It can also be observed from the figures that both the tools (dedicated code [110, 139] and Cascade 3.0 scattering matrix code [140]) have been resulted in the similar responses for both the designs.

7.3.2 Design of a Double Disc RF-Window for a 170 GHz Gyrotron

In this subsection, the design of a double disc RF-window is presented for a 170 GHz, 1 MW, CW gyrotron (Gaussian mode) to be used in a Electron Cyclotron Resonance Heating (ECRH) applications [129, 131]. A CVD-diamond disc material was considered in this design. The schematic diagram for this

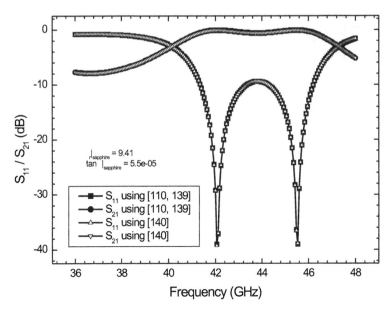

Fig. 7.4 Optimized transmission and reflection characteristics of a double-disc window with Sapphire disc material for a 42 GHz gyrotron ($D_1=D_2=2.28$ mm, $t_{dd}=3.6$ mm, $r_{dd}=41.2$ mm) [132] (© 2009 IEEE)

disc is same as Fig. 7.1. The design parameters considered in this example and their range are specified in Table 7.3. The disc thickness for Gaussian mode is given by [110, 129],

$$t_{dd} = N\frac{\lambda_n}{2} = N\frac{c}{2f\sqrt{\epsilon_r}} \tag{7.4}$$

where f is the desired resonant frequency and ϵ_r represents dielectric property of the disc material. The objective functions considered for this design are same as those given in Eq.s (7.2) and (7.3). The analysis of the window during the iterative optimization process was carried out following the same dedicated code [110, 139] mentioned in the previous subsection.

 The Pareto front for a double disc RF-window for 170 GHz gyrotron is shown in Fig. 7.6. One of the best optimized responses from the Pareto front is selected, and it is shown in Fig. 7.7 along with the values of optimized design parameters. It can be observed that the best design from the Pareto front of double disc window (for 170 GHz) resulted in exact matching of desired frequency with a large bandwidth of 3.7 GHz measured at −30 dB level.

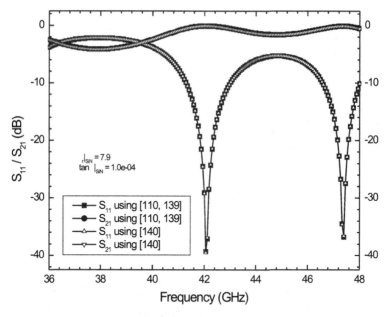

Fig. 7.5 Optimized transmission and reflection characteristics of a double-disc window with SiN disc material for a 42 GHz gyrotron ($D_1=D_2=2.53$ mm, $t_{dd}=3.02$ mm, $r_{dd}=40.0$ mm) [132] (© 2009 IEEE)

7.3.3 Design of a Pillbox-Type RF-Window for a 2.856 GHz Klystron

As a third design problem, the design of a pillbox-type RF-window for 2.856 GHz 5 MW, pulsed klystron is presented. The schematic diagram of a pillbox-type RF-window is shown in Fig. 7.8. Three design parameters, namely, length of the pillbox window (L_{pb}), the thickness of the disc (t_{pb}), and the diameter of the cylindrical disc (d_{pb}) are considered for optimizing the performance of the window. The range of these parameters considered for optimization is specified in Table 7.4. In this experiment, the range of the design parameters are determined from a specific design of pillbox-type window given in [130, 138]. The objective functions considered for getting the Pareto front are same as those used in double disc design experiments. The BW in objective function f_2 was measured at -30 dB level from the S-Parameters curve in this experiment.

The Pareto front comprising optimized solutions of a pillbox-type RF-window is shown in Fig. 7.9. One of the best optimized solution from the Pareto front is selected, and its response is shown in Fig. 7.10. It can be observed from the figure that the optimized design of pillbox-type RF-window has resulted in matching at desired frequency with a bandwidth of 475 MHz measured at -30 dB level around the resonant frequency.

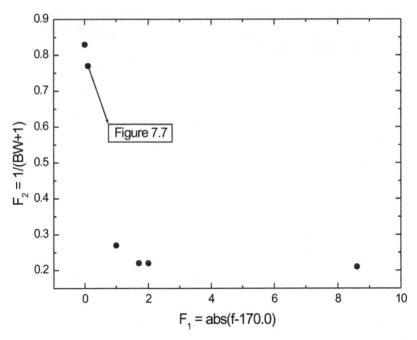

Fig. 7.6 Pareto front of a double disc RF-window for a 170 GHz gyrotron [132] (© 2009 IEEE)

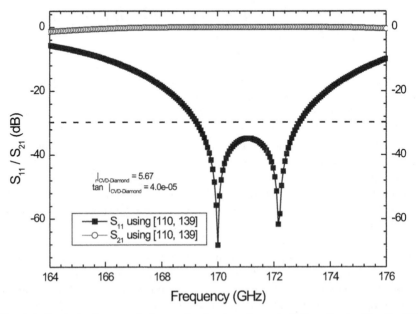

Fig. 7.7 Optimized transmission and reflection characteristics of a double-disc RF-window for a 170 GHz gyrotron ($D_1=D_2=1.111$ mm, $t_{dd}=3.0$ mm, $r_{dd}=55.0$ mm) [132] (© 2009 IEEE)

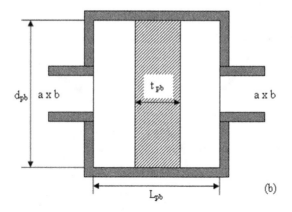

Fig. 7.8 Schematic diagram of a pillbox-type RF-window (a=72.14 mm, and b= 34.04 mm [130]).

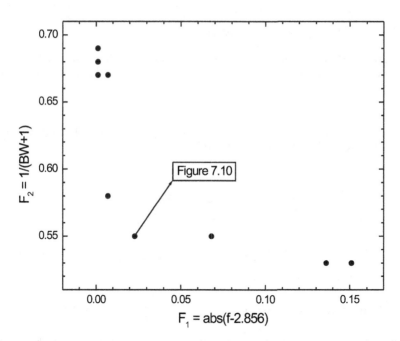

Fig. 7.9 Pareto front of a pillbox-type RF-window for a 2.856 GHz klystron [132] (© 2009 IEEE)

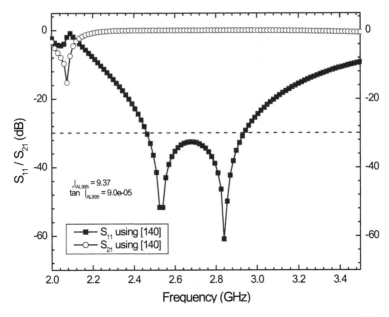

Fig. 7.10 Optimized transmission and reflection characteristics of a pillbox-type RF-window for a 2.856 GHz klystron (L_{pb}=37.07 mm, t_{pb}=2.86 mm, d_{pb}=88.1 mm) [132] (© 2009 IEEE)

Table 7.4 Range of design parameters for a pillbox-type RF-window for a 2.856 GHz Klystron [132, 133] (© 2009 IEEE)

Design Parameters	Range [mm]
Length (L_{pb})	35.0-40.0
Disc thickness (t_{pb})	2.0-4.0
Diameter (D_{pb})	87.0-92.0

7.4 Concluding Remarks

In this chapter, a multi-objective optimization method with particle swarm optimizer is demonstrated for the design of disc-type RF-windows. The optimum trade-off between the objectives of matching with desired resonant frequency and minimizing the reflections around the resonant frequency (by maximizing BW around it) is achieved using the concept of MOPSO. The design and optimization of three RF-windows - a double disc window for 42 GHz gyrotron (with Sapphire and SiN as disc materials), a double disc window for 170 GHz gyrotron, and a pillbox-type window for 2.856 GHz klystron - are obtained as case studies of high power microwave and millimeter-wave components. In all experiments, the MOPSO algorithm converged with diversities in the solutions of Pareto optimal set.

Here, the solution for the design of RF-windows can also become possible by formulating single objective. In this case, the matching at desired frequency should be treated as a constraint, while minimizing the reflections around resonant frequency becomes the objective function. For many microwave and millimeter-wave design problems, more specifically for the problems with high working frequencies (e.g. double disc windows at 42 GHz and 170 GHz gyrotrons used in presented experiments), fixing of the desired frequency along with other objectives is quite difficult. This is one of the reason why in this problem the matching of the desired frequency is treated as a separate objective rather than making it a constraint. Treating both as different objectives, MOO methods find optimum trade-off between them. Another advantage in designing components using MOO methods is that it gives a set of solutions instead of a single solution which is obtained in single objective optimization. This gives the designer a choice for fixing (selecting) design parameters from a set of optimized solutions based on the compromise between the available objectives and the requirement of the problem.

However, in all our experiments the disc diameter (or radius) is considered as an important design parameter, it can also be fixed depending upon the material selection and availability. In all, it can be concluded that the use of multi-objective particle swarm optimization is an efficient optimization tool for tuning design parameters and getting optimum response of a wide range of high power millimeter-wave devices/components, and it can also be explored for other critical high power microwave and millimeter-wave design applications.

8

Concluding Remarks

In this chapter, the contents presented in the book are summarized, and further readings and experiments are suggested.

8.1 Summary of the Book

In this book various modified and hybrid soft computing methods are presented and used for diverse microwave design applications as case studies. The main object behind this exercise is to make the design as fast as possible while improving the quality/accuracy of the design. This is demonstrated using applications such as design of microwave filter, microstrip antenna, and modeling of one-port microstrip via. The book also deals with the design of critical applications where high precision with little tolerance is the prime requirement. Here, we considered the design of two important components, namely, a nonlinear taper and RF-windows for high power microwave sources.

The detailed contributions of the book are as follows:

- A novel modification to the Particle Swarm Optimization (PSO) algorithm is presented in the book and its applicability is demonstrated for the design of a specific microwave filter as a case study of microwave components. In the modified approach, a novel paradigm of multiple subswarms for searching parameter space with PSO algorithm is introduced. The particles in the swarm are divided to form multiple subswarms. The social component of PSO's velocity update equation is modified to include the effects of multiple subswarms. The presented approach, PSO with Multiple Subswarms (PSO-MS), is tested with a test-bed of five benchmark functions which are commonly used to compare the performance of different Evolutionary Algorithms (EAs) and their modifications. The concept is tested by applying it to two basic PSO variations, namely, PSO with Inertia Weight Method (IWM) and PSO with Constriction Factor Method (CFM). We have observed from the results obtained in Chapter 3, that if computational time is the prime requirement, we can fix the number of particles constant and

N. Chauhan, M. Kartikeyan, and A. Mittal: Soft Computing Methods, SCI 392, pp. 97–100.
springerlink.com   © Springer-Verlag Berlin Heidelberg 2012

achieve faster convergence, whereas if accuracy is the critical requirement the number of particles should also be increased in accordance with the number of subswarms. Finally, the presented PSO-MS algorithm is used for the design of coupled microstrip-line band pass filter. This is a computationally expensive problem when the design is conducted using EAs that invoke EM simulators in the optimization loop. The design results obtained with the PSO-MS algorithm are compared with those obtained using standard PSO. The comparison of results for the microstrip filter design problem proves the applicability of proposed approach in other microwave design problems where computational time is an important factor. Another advantage of this algorithm is that it can be parallelized easily as each subswarm's computation can be done on a separate processor. As a whole, the goal of obtaining faster design while improving quality of the design is fulfilled.

- An effective method for modeling microwave components using Support Vector Machine (SVM) is presented in the book. Modeling of three microwave components, namely, a one-port GaAs microstrip via, a circular polarized microstrip antenna and an aperture coupled microstrip antenna are carried out using the presented framework. The process of data generation, scaling, kernel selection, and parameter selection is described. The models developed using SVM are compared with those developed using existing ANN method. The results obtained indicate that the SVM model is better compared to that of ANN model due to its higher accuracy and generalization ability.

- A hybrid approach combining SVM with EAs is presented for microwave design applications. The aim behind the approach is to reduce the large computational expenses incurred by EM simulations when they are used in optimization loops of EAs. In this approach, an approximate model of the EM simulation based design process for the component to be designed is created using SVM. To obtain an optimum combination of design parameters, this approximate model (instead of EM simulators) is invoked in the optimization loop of EAs to get the output response. The hybrid approach is named as Support Vector driven Evolutionary Algorithms (SVEA). The exciting advantage obtained by the SVM model is that it responds quickly (approximately in milliseconds) compared to the parametric analysis of EM simulations which respond approximately in minutes depending upon the complexity of the component's structure. In this work, two different EAs - GA and PSO are used to optimize the model generated with SVM, making the approaches Support Vector driven Genetic Algorithms (SVGA) and Support Vector driven Particle Swarm Optimization (SVPSO). Both the hybrid approaches (SVGA and SVPSO) are used for the design of two microstrip antennas - a circular polarized microstrip antenna and an aperture coupled microstrip antenna respectively. The optimized design of circular polarized microstrip antenna obtained using the hybrid SVGA approach is also compared with the optimized design obtained using similar hybrid approach - a Neural Network driven Genetic Algorithms (NNGA), in which the approximate model is developed using well-known ANN method. The approach presented here is found to be highly effective in reducing computational expenses incurred by EM simulators.

- In this book, a Modified Bacterial Foraging Optimization (MBFO) algorithm is presented which includes three changes to the standard BFO technique. First, the bacteria are implemented with memory so that whenever bacteria encounters unfavorable environment, they return to their previous memorized position, tumble and swim again in the chemotaxis loop. Second, the PSO operator with only social component is applied after each chemotaxis step [79] to improve the convergence ability of the algorithm. Third, variable swim length is used to facilitate the bacterium to reach global optima faster. A comparison of presented MBFO algorithm is made with state-of-the-art version of standard PSO algorithm using five benchmark functions which proved the effectiveness of the presented algorithm.

- In this book, the design of a nonlinear taper for a 42 GHz, 200 kW, CW gyrotron is presented. This work is done as part of a project entitled *"Design and development of 42 GHz, 200 kW, CW, long pulse gyrotron"*, which is presently undertaken in India. Our contribution in this work is to carry out the design optimization of a nonlinear taper to be used in the output system of the gyrotron. The power transmission requirement for this design is above 99%. This is a critical design application in which high precision and accuracy are required. The design optimization of a raised cosine type nonlinear taper is carried out using two swarm intelligence based algorithms, namely, PSO and MBFO. The parametric analysis of the taper is carried out for the selected design parameters and it is identified that the parameter gamma (γ) which is used in the synthesis of raised cosine taper profile plays critical role in obtaining desired response. The analysis of the taper was carried out using a dedicated Scattering Matrix Code (SMC) [125]. The optimization parameters obtained using both the above mentioned algorithms resulted in an excellent transmission of 99.87% in the desired $TE_{0,3}$ mode with minimum spurious mode generation. It is concluded that a state-of-the-art version of PSO and MBFO can be used for the design of various other high power microwave and millimeter wave components.

- The concept of multi-objective optimizations using PSO algorithm is demonstrated considering by the design of disc-type RF-windows. The RF-window is also a critical output component in high power microwave and millimeter wave sources and needs careful design. The design and optimization of three RF-windows - a double disc RF-window for 42 GHz, 200 kW, CW gyroton (with sapphire and SiN disc materials), a double disc RF-window for a 170 GHz, 1 MW, CW gyrotron, and a pillbox-type RF-window for a 2.856 GHz, 5 MW, pulsed klystron - are presented. Due to sensitivity in matching with desired frequency while minimizing the reflections, a Multi-objective Particle Swarm Optimization (MOPSO) approach is chosen in this design. To carry out the design process, a specific implementation, MOPSO with crowding distance [136], is used. An optimum trade-off between the objectives of matching desired resonant frequency and minimizing the reflections around the resonant frequency (by maximizing BW) is achieved. It is demonstrated that a well distributed Pareto front is obtained in all the three experiments. It is concluded that MOPSO method can be useful for the design of high power microwave and millimeter wave sources when there are multiple objectives to be optimized.

8.2   Further Enhancements

There are a number of useful extensions that can be added to the the present research work. A few soft computing methods have been developed and used for solving challenging design problems in microwave domain. Yet the other design problems with the same or different challenges can also be picked up and solutions be found with the current or other emerging soft computing methods. The possible additions to the present work are stated below:

- It is possible to use soft computing methods presented in this book for the other microwave design applications with similar challenges. It is also possible to apply these methods for the design problems of other engineering disciplines.
- The implementation of EAs such as GA, PSO, BFO, and their modifications on multi-core environments is a promising research area. This can be helpful in performing faster design of microwave components. A parallel implementation of PSO on multi-core environment has been tested on test problems by the authors [142]. The similar kind of parallel implementation may be used for design problems of microwave and millimeter-wave field in order to fasten the process.
- Parameter selection of SVM is also an open research issue. As the values of the critical hyperparameters affect the accuracy of the SVM model, an effective method may be developed for the optimum selection of these parameters.
- Due to continuous advancements in soft computing methods, it is possible to investigate other emerging methods such as artificial immune system, space mapping, differential evolution, ant colony optimization, etc., for microwave design problems. It may be useful to compare the performances of these methods.
- It is also possible to perform improvements and hybridizations to the present and emerging soft computing methods in order to remove the drawbacks and exploit the merits of these methods.
- As there are many user defined parameters of BFO algorithm, it is possible to carry out parametric analysis of BFO algorithm. It may also be possible to develop a method for the reduction of dependency on user defined parameter selection for BFO and also in other EAs.

9

Glossary

Antenna: It is a metallic structure used to transmit and/or receive radio waves.

Aperture coupled microstrip antenna: It is a microstrip antenna in which the power coupling between the patch and the feedline is done via the aperture (slot) on the ground plane.

Axial ratio: It an important antenna parameter that is defined as the ratio of major axis to minor axis of the field polarization.

Axial ratio bandwidth: It is the frequency range over which the antenna exhibits axial ratio parameter to be at a specified level (which is generally taken less than or equal to 3 dB for circular polarization).

Bandwidth: Bandwidth is the frequency range within which the component meets desired specification (say gain, impedance or VSWR).

Circular polarization: When a field vector at a given point in space rotates with a same magnitude and a phase shift of 90° is said to be circularly polarized.

Cross validation: It is a procedure to estimate the performance of a classifier/regressor. In a v-fold cross validation, training set is divided into v-subsets of equal size. Repeatedly each subset is tested using the classifier/regressor which is trained on remaining v-1 subsets.

Crowding distance: It is a metric that determines density of solutions around a particular solution.

Evolutionary algorithm: It is a subset of evolutionary computation that is implemented as evolutionary process to form optimization algorithm.

Fitness: The fitness is a measure of how well a particular solution performs at solving the problem.

Fitness function: It is an objective function that determines fitness of a solution.

Foraging theory: It is a theory based on foraging behavior of biological population (animals) that search for nutrients to maximize their energy intake per unit time.

N. Chauhan, M. Kartikeyan, and A. Mittal: Soft Computing Methods, SCI 392, pp. 101–102.
springerlink.com   © Springer-Verlag Berlin Heidelberg 2012

Gyrotron: It is an electronic device capable of generating high power at microwave and millimetric wavelengths.

Klystron: It is a widely used vacuum tube as a generator or as a microwave amplifier.

Microstrip antenna: It is a passive component that consists of a radiating element (patch) mounted over a grounded dielectric substrate.

Microwave filter: It is a two port network to control frequency response in microwave systems.

Microstrip via: It is a component used to interconnect micrstrip circuits.

Pareto front: The set of objective function vectors corresponding to Pareto-optimal set in the objective space is referred to as Pareto front.

Pareto-optimal set: A set of non-dominated solutions with respect to given objectives is referred to as Pareto-optimal set. A solution is non-dominated if its one objective can not be improved without loss of one or more objectives.

Polarization: Polarization is a property of an electromagnetic wave describing the time varying direction and relative magnitude of the field vector at a given point in space.

RF-window: It is a component used in the output system of high power microwave/millimeter-wave devices for power transition from the device to transmission line and vice versa.

Scattering matrix: It is a linear relationship between input and output that involves precisely measurable parameters which are known as S-parameters. S-parameters such as S_{11} (reflection coefficient) and S_{21} (transmission coefficient) are important for measuring transmission and reflection characteristics of a two port network.

Spurious modes: The modes other than the desired mode are spurious modes. The design should be carried out in such a way that the spurious mode contents are reduced.

Swarm intelligence: It is a problem solving behavior that emerge from a group (swarm) of agents which communicate among each other based on their local environment.

Tapered transmission line (Taper): It is a transmission line section which is used to connect transmission lines of different cross sectional areas.

Voltage Standing Wave Ratio (VSWR): It is a ratio of the maximum voltage to minimum voltage (i.e., V_{max}/V_{min}) due to mismatch of the load in the transmission line.

References

1. Deb, K.: Optimization for engineering design algorithms and examples. Prentice-Hall India, New Delhi (2003)
2. Fletcher, R.: Practical methods of optimization, 2nd edn. John Wiley & Sons (1987)
3. A Definition of Soft Computing - adapted from L. A. Zadeh, http://www.soft-computing.de/def.html (Last accessed on September 2011)
4. Pratihar, D.K.: Soft computing. Narosa Publishing House, New Delhi (2008)
5. Zadeh, L.A.: Forward. In: Proceedings of the Second International Conference on Fuzzy Logic and Neural Networks, Iizuka, Japan, pp. XIII-XIV(1992)
6. Zhang, Q.J., Gupta, K.C.: Neural networks for RF and microwave design. Artech House, Boston (2000)
7. Rahmat-Samii, Y., Michielssen, E. (eds.): Electromagnetic optimization by genetic algorithms. John Wiley & Sons, New York (1999)
8. Weile, D.S., Michielssen, E.: Genetic algorithm optimizations applied to electromagnetics: a review. IEEE Transactions on Antenna and Propagation 45(3), 343–353 (1997)
9. Linden, D.S., Altshuler, E.E.: Evolving wire antennas using genetic algorithms: a review. In: Proceedings of 1st NASA/DOD Workshop on Evolvable Hardware (EH 1999), Pasadena, CA, USA, July 19-21, pp. 225–232 (1999)
10. Haupt, R.L., Werner, D.H.: Genetic algorithms in electromagnetics. Wiley-IEEE Press, Hoboken (2007)
11. Johnson, J.M., Rahmat-Samii, Y.: Genetic Algorithms and Method of Moments (GA/MoM) for design of integrated antennas. IEEE Transactions on Antenna and Propagation 47(10), 1606–1614 (1999)
12. Hong, Y., Dong, Z.: Genetic algorithms with applications in wireless communications. International Journal of Systems Science 35(13), 751–762 (2004)
13. Villegas, F., Cwik, T., Rahmat-Samii, Y., Manteghi, M.: A parallel Electromagnetic Genetic-algorithm Optimization (EGO) application for patch antenna design. IEEE Transactions on Antennas and Propagation 52(9), 2424–2435 (2004)
14. Yan, K., Lu, Y.: Sidelobe reduction in array-pattern synthesis using genetic algorithm. IEEE Transactions on Antennas and Propagation 45(7), 1117–1122 (1997)

15. Plaum, B., Wagner, D., Kasparek, W., Thumm, M.: Optimization of oversized waveguide components using genetic algorithm. Fusion Engineering and Design 53, 499–503 (2001)
16. Wang, F., Zhang, Q.J.: Knowledge-based neural models for microwave design. IEEE Transactions on Microwave Theory and Techniques 45(12), 2333–2343 (1997)
17. Marinova, I., Panchev, C., Katsakos, D.: A neural network inverse approach to electromagnetic device design. IEEE Transactions on Magnetics 36(4), 1080–1084 (2000)
18. Suntives, A., Hossain, M.S., Ma, J., Mittra, R., Veremey, V.: Application of artificial neural network models to linear and nonlinear RF circuit modeling. International Journal of RF and Microwave Computer-Aided Engineering 11, 231–247 (2001)
19. Ding, X., Devabhaktuni, V.K., Chattaraj, B., Yagoub, M.C.E., Deo, M., Xu, J., Zhang, Q.J.: Neural-network approaches to electromagnetic-based modeling of passive components and their applications to high-frequency and high-speed nonlinear circuit optimization. IEEE Transactions on Microwave Theory and Techniques 52(1), 436–449 (2004)
20. Kabir, H., Wang, Y., Yu, M., Zhang, Q.J.: Applications of artificial neural network techniques in microwave filter modeling, optimization and design. Progress In Electromagnetic Research 3(7), 1131–1135 (2007)
21. Mishra, S., Bhende, C.N.: Bacterial foraging technique-based optimized active power filter for load compensation. IEEE Transactions on Power Delivery 22(1), 457–465 (2007)
22. Nair, S.B., Kumar, A.: An Artificial Immune System Based Approach for English Grammar Checking. In: de Castro, L.N., Von Zuben, F.J., Knidel, H. (eds.) ICARIS 2007. LNCS, vol. 4628, pp. 348–357. Springer, Heidelberg (2007)
23. Chauhan, N.C.: Soft computing methods for design applications in microwave domain. PhD Thesis, Indian Institute of Technology Roorkee, Roorkee, India (2010)
24. Holland, J.H.: Adaptation in natural and artificial systems. The University of Michigan Press, Ann Arbor (1975)
25. Goldberg, D.E.: Genetic algorithm in search, optimization and machine learning. Pearson Education, Singapore (1989)
26. Rajasekaran, S., Pai, G.A.V.: Neural networks, fuzzy logic, and genetic algorithms - synthesis and applications. Prentice-Hall of India, New Delhi (2006)
27. Eberhart, R., Kennedy, J.: A new optimizer using particle swarm theory. In: Proceedings of 6th International Symposium Micro Machine and Human Science (MHS 1995), Nagoya, Japan, October 4-6, pp. 39–43 (1995)
28. Hu, X., Eberhart, R.C., Shi, Y.: Engineering optimization with particle swarm. In: Swarm Intellignece Symposium (SIS-2003), Indiana, USA, April 24-26, pp. 53–57 (2003)
29. Engelbrecht, A.P.: Fundamentals of computational swarm intelligence. John Wiley & Sons, Hoboken (2005)
30. Shi, Y., Eberhart, R.C.: A modified particle swarm optimizer. In: Proceedings of IEEE World Congress on Computational Intelligence Evolutionary Computation, Anchorage, USA, May 4-9, pp. 69–73 (1998)

31. Shi, Y., Eberhart, R.C.: Empirical study of particle swarm optimization. In: Proceedings of the Congress on Evolutionary Computation, vol. 3, pp. 1945–1950 (1999)

32. Clerc, M.: The swarm and the queen: towards a deterministic and adaptive particle swarm optimization. In: Proceedings of IEEE Congress on Evolutionary Computation(CEC 1999), Washington, DC, USA, pp. 1951–1957 (1999)

33. Eberhart, R.C., Shi, Y.: Comparing inertia weights and constriction factors in particle swarm optimization. In: Proceedings of IEEE Congress on Evolutionary Computation (CEC 2000), La Jolla, USA, vol. 1, pp. 84–88 (July 2000)

34. Passino, K.M.: Biomimicry of bacterial foraging for distributed optimization and control. IEEE Control Systems Magazine 22, 55–67 (2002)

35. Haykin, S.: Neural networks - a comprehensive foundation. Prentice Hall of India, New Delhi (2004)

36. Ham, F.M., Kostanic, I.: Principles of neurocomputing for science and engineering. Tata McGraw-Hill, New Delhi (2002)

37. Graphical structure of biological neuron, http://www.neuralpower.com/technology.htm (Last accessed on September 2011)

38. Vapnik, V.: The nature of statistical learning theory. Springer, New York (1995)

39. Cortes, C., Vapnik, V.: Support vector networks. Machine Learning 20, 273–297 (1995)

40. Gunn, S.R.: Support vector machines for classification and regression. Technical Report, Image Speech and Intelligent Systems Research Group, University of Southampton (1997)

41. Cristianini, N., Shawe-Taylor, J.: An introduction to support vector machines and other kernel-based learning methods. Cambridge University Press, Cambridge (2000)

42. Smola, A., Schölkopf, B.: A tutorial on support vector regression. Neuro-COLT2 Technical Report Series, NC2-TR-1998-030 (1998)

43. Schölkopf, B., Mika, S., Burges, C.J.C., Knirsch, P., Muller, K.R., Ratsch, G., Smola, A.J.: Input space versus feature space in kernel-based methods. IEEE Transactions on Neural Networks 10(5), 1000–1017 (1999)

44. Chauhan, N.C., Kartikeyan, M.V., Mittal, A.: A review on the use of soft computing methods for microwave design applications. FREQUENZ 63(1-2), 24–31 (2009)

45. Mahanti, G., Pathak, N., Mahanti, P.K.: Synthesis of thinned linear antenna arrays with fixed sidelobe level using real-coded genetic algorithm. Progress In Electromagnetics Research 75, 319–328 (2007)

46. Robinson, J., Rahmat-Samii, Y.: Particle swarm optimization in electromagentics. IEEE Transactions on Antennas and Propagation 52(2), 397–407 (2004)

47. Ciuprina, G., Ioan, D., Munteanu, I.: Use of intelligent-particle swarm optimization in electromagnetics. IEEE Transactions on Magnetics 38(2), 1037–1040 (2002)

48. Jin, N., Rahmat-Samii, Y.: Particle swarm optimization for antenna designs in engineering electromagnetics. Journal of Artificial Evolution and Applications, Hindawi Publishing Corporation, Article ID 728929, 10 pages (February 2008)

49. Jin, N.B., Rahmat-Samii, Y.: Parallel Particle Swarm Optimization and Finite-Difference Time-Domain (PSO/FDTD) algorithm for multiband and wide-band patch antenna designs. IEEE Transactions on Antennas and Propagation 53(11), 3459–3468 (2005)

50. Xu, S., Rahmat-Samii, Y.: Multi-objective particle swarm optimization for high performance array and reflector antennas. In: Proceedings of IEEE Antenna and Propagation Society International Symposium, Albuquerque, Mexico, July 9-14, pp. 3293–3296 (2006)

51. Wang, W., Lu, Y., Fu, J.S., Xiong, Y.Z.: Particle swarm optimization and finite-element based approach for microwave filter design. IEEE Transactions on Magnetics 41(5), 1800–1803 (2005)

52. Mikki, S.M., Kishk, A.A.: Particle swarm optimization: a physics-based approach. In: Synthesis Lectures on Computational Electromagnetics, Morgan & Claypool (2008)

53. Kennedy, J., Eberhart, R.C.: Particle swarm optimization. In: Proceedings of IEEE International Conference on Neural Networks, Perth, Australia, vol. 4, pp. 1942–1948 (November 1995)

54. Hassan, R., Cohanim, B., Weck, O.: A comparison of particle swarm optimization and the genetic algorithm. In. In: Proceedings of 46th AIAA/ASNE/ASCE/AHS/ASC Structures, Structural Dynamics and Material Conference, Austin, TX, April 18-21 (2005)

55. Boeringer, D.W., Werner, D.H.: Particle swarm optimization vs. genetic algorithms for phased array synthesis. IEEE Transactions on Antenna and Propagation 52(3), 771–779 (2004)

56. Gollapudi, S.V.R.S., Pattnaik, S.S., Bajpai, O.P., Devi, S., Sagar, C.V., Pradymna, P.K., Bakwad, K.M.: Bacterial foraging optimization technique to calculate resonant frequency of rectangular microstrip antenna. International Journal of RF and Microwave Computer-Aided Engineering 18, 383–388 (2008)

57. Datta, T., Misra, I.S., Mangaraj, B.B., Imtiaj, S.: Improved adaptive bacteria foraging algorithm in optimization of antenna array for faster convergence. Progress In Electromagnetics Research C 1, 143–157 (2008)

58. Guney, K., Basbug, S.: Interference supression of linear antenna arrays by amplitude-only control using a bacterial foraging algorithm. Progress In Electromagnetics Research 79, 475–497 (2008)

59. Wu, C., Zhang, N., Jiang, J., Yang, J., Liang, Y.: Improved Bacterial Foraging Algorithms and their Applications to Job Shop Scheduling Problems. In: Beliczynski, B., Dzielinski, A., Iwanowski, M., Ribeiro, B. (eds.) ICANNGA 2007. LNCS, vol. 4431, pp. 562–569. Springer, Heidelberg (2007)

60. Majhi, R., Panda, G., Dash, P.K., Das, D.P.: Stock market prediction of S & P 500 and DJIA using bacterial foraging optimization technique. In: IEEE Congress on Evolutionary Computation (CEC 2007), September 25-28, pp. 2569-2575, Singapore (2007)

61. Coelho, L.S., Silveira, C.C.: Improved bacterial foraging strategy for controller optimization applied to robotic manipulator system. In: Proceedings of the 2006 IEEE International Symposium on Intelligent Control, Munich, Germany, pp. 1276–1281 (October 4-6, 2006)

62. Chauhan, N.C., Kartikeyan, M.V.: Application of soft computing techniques for the design optimization of specific components of high power gyrotrons. In: 5th IAEA Technical Meeting on ECRH Physics and Technology for Large Fusion Devices, February18-20, ITER-India, Institute for Plasma Research, Gandhinagar (2009)

63. Clarke, S.M., Griebsch, J.H., Simpson, T.W.: Analysis of support vector regression for approximation of complex engineering analysis. ASME Transactions on Mechanical Design 127, 1077–1087 (2005)

64. Burrascano, P., Fiori, S., Mongiardo, M.: A review of artificial neural networks applications in microwave computer-aided design. International Journal of RF and Microwave Computer-Aided Engineering 9, 158–174 (1999)

65. Angiulli, G., Cacciola, M., Versaci, M.: Microwave devices and antennas modeling by support vector regression machines. IEEE Transactions on Magnetics 43(4), 1589–1592 (2007)

66. Angiulli, G., Barrile, V., Cacciola, M.: Solving electromagnetic inverse scattering problems by SVRMs: a case of study towards georadar applications. PIERS Online 3(5), 741–745 (2007)

67. Wu, Y., Tang, Z., Xu, Y., Guo, Y., Zhang, B.: Support vector regression for measuring electromagnetic parameters of magnetic thin-film materials. IEEE Transactions on Magnetics 43(12), 4071–4075 (2007)

68. Güneş, F., Türker, N., Gürgen, F.: Support vector design of microstrip lines. International Journal of RF and Microwave Computer-Aided Engineering 18, 326–336 (2008)

69. Günes, F., Türker, N., Gürgen, F.: Signal-noise support vector model of a microwave transistor. International Journal of RF and Microwave Computer-Aided Engineering 17, 404–415 (2007)

70. Martínez-Ramón, M., Christodoulou, C.: Support vector machines for antenna array processing and electromagnetics. In: Synthesis Lectures on Computational Electromagnetics, Morgan & Claypool Publishers (2006)

71. Xu, Y., Guo, Y., Xu, R., Xia, L., Wu, Y.: A support vector regression based nonlinear modeling method for SiC MESFET. Progress In Electromagnetics Research Letters 2, 103–114 (2008)

72. Kumar, G.S., Kalra, P.K., Dhande, S.G.: Hybrid computation using neuro-genetic and classical optimization for B-spline curve and surface fitting. International Journal of Hybrid Intelligent Systems 1(4), 176–188 (2004)

73. Jadon, R.S., Chaudhury, S., Biswas, K.K.: Generic video classification: an evolutionary learning based fuzzy theoretic approach. In: Proceedings of the Third Indian Conference on Computer Vision, Graphics & Image Processing (ICVGIP 2002), December 16-18, Space Applications Centre, Ahmadabad (2002)

74. Abraham, A., Philip, S., Mahanti, P.K.: Soft computing models for weather forecasting. International Journal of Applied Science and Computation 11(3), 106–117 (2004)

75. Chaturvedi, D.K., Kumar, R., Kalra, P.K.: Artificial neural network learning using improved genetic algorithms. Journal of the Institution of Engineers (India) 82, 1–8 (2001)

76. Yang, Y., Chen, R.S., Ye, Z.B.: Combination of particle swarm optimization with least-square support vector machine for FDTD time series forecasting. Microwave and Optical Technology Letters 48(1), 141–144 (2006)

77. Robinson, J., Sinton, S., Rahmat-Samii, Y.: Particle swarm, genetic algorithm, and their hybrids: optimization of a profiled corrugated horn antenna. In: IEEE Antenna and Propagation Society International Symposium, San Antonio, TX, USA, vol. 1, June 16-21, pp. 314–317 (2002)

78. Gandelli, A., Grimaccia, F., Mussetta, M., Pirinoli, P., Zich, R.E.: Genetical swarm optimization: an evolutionary algorithm for antenna design. ATKAAF 47(3-4), 105–112 (2006)

79. Biswas, A., Dasgupta, S., Das, S., Abraham, A.: Synergy of PSO and Bacterial Foraging Optimization - a Comparative Study on Numerical Benchmarks. In: Corchado, E., et al. (eds.) Innovations in Hybrid Intelligent Systems. ASC, vol. 44, pp. 255–263. Springer, Heidelberg (2007)

80. Kim, D.H., Abraham, A., Cho, J.H.: A hybrid genetic algorithm and bacterial foraging approach for global optimization. International Journal of Information Sciences 177, 3918–3937 (2007)

81. Binkley, K.J., Hagiwara, M.: Particle swarm optimization with area of influence: increasing the effectiveness of the swarm. In: Proceedings of IEEE Swarm Intelligence Symposium (SIS-2005), Pasadena, California, USA, June 8-10, pp. 45–52 (2005)

82. Chauhan, N.C., Kartikeyan, M.V., Mittal, A.: A modified particle swarm optimizer and its application to the design of microwave filters. Journal of Infrared, Millimeter and Terahertz Waves 30(6), 598–610 (2009)

83. Shi, Y., Eberhart, R.C.: Parameter Selection in Particle Swarm Optimization. In: Porto, V.W., Waagen, D. (eds.) EP 1998. LNCS, vol. 1447, pp. 591–601. Springer, Heidelberg (1998)

84. Zhang, L., Yu, H., Hu, S.: Optimal choice of parameters for particle swarm optimization. Journal of Zhejiang University Science 6A(6), 528–534 (2005)

85. Ratnaweera, A., Halgamuge, S.K., Watson, H.C.: Self-organizing hierarchical particle swarm optimizer with time-varying acceleration coefficients. IEEE Transactions on Evolutionary Computation 8(3), 240–255 (2004)

86. Matthaei, G., Young, L., Jones, E.M.T.: Microwave filters, impedance-matching networks, and coupling structures. Artech House, MA (1980)

87. Bhatti, R.A., Kayani, J.K.: Design and analysis of a parallel coupled microstrip band pass filter. In: Proceedings of 2nd International Bhurban Conference on Applied Science and Technology, Bhurban, Pakistan, pp. 16–21 (June 2003)

88. Pozar, D.M.: Microwave engineering, 3rd edn. John Wiley & Sons, New York (2005)

89. Levy, R., Snyder, R.V., Matthaei, G.: Design of microwave filters. IEEE Transactions on Microwave Theory and Techniques 50(3), 783–793 (2002)

90. Mandal, M.K., Sanyal, S.: Design of wide-band, sharp-rejection bandpass filters with parallel-coupled lines. IEEE Microwave and Wireless Components Letters 16(11), 597–599 (2006)

91. IE3D: MoM-Based EM Simulator, Zeland's Software Inc., Release 10.2 (2003)

92. Wu, D.I., Chang, D.C., Brim, B.L.: Accurate numerical modeling of microstrip junctions and discontinuities. International Journal of Microwave and Millimeter-wave Computer-Aided Engineering (MIMICAE) 1(1), 48–58 (1991)

93. Montgomery, D.C.: Design and analysis of experiments, 4th edn. John Wiley & Sons, New York (1997)

94. Bandler, J.W., Biernacki, R.M., Chen, S.H., Grobelny, P.A., Hemmers, R.H.: Space mapping technique for electromagnetic optimization. IEEE Transactions on Microwave Theory and Techniques 42(12), 2536–2544 (1994)

95. Watson, P.M., Gupta, K.C.: EM-ANN models for microstrip vias and interconnects in dataset circuits. IEEE Transactions on Microwave Theory and Technique 44(12), 2495–2503 (1996)

96. Devabhaktuni, V.K., Yagoub, M.C.E., Fang, Y., Xu, J., Zhang, Q.J.: Neural networks for microwave modeling: model development issues and nonlinear modeling techniques (Invited review). International Journal of RF and Microwave Computer Aided-Engineering 11, 4–21 (2001)

97. Chauhan, N.C., Roy, Y.K., Kumar, A., Kartikeyan, M.V., Mittal, A.: SVM-PSO based modeling and optimization of microwave components. FREQUENZ 62(1-2), 18–24 (2008)

98. Chauhan, N.C., Kartikeyan, M.V., Mittal, A.: Support vector driven genetic algorithm for the design of circular polarized microstrip antenna. International Journal of Infrared, Millimeter and Terahertz Waves 29(6), 558–569 (2008)

99. Shinghal, R.: Pattern recognition - techniques and applications. Oxford University Press, New Delhi (2006)

100. Keerthi, S.S.: Efficient tuning of SVM hyperparameters using radius/margin bound and iterative algorithms. IEEE Transactions on Neural Networks 13(5), 1225–1229 (2002)

101. Kumar, A., Kartikeyan, M.V.: A circular polarized stacked patch aperture coupled microstrip antenna for 2.6 GHz band. International Journal of Infrared Millimeter Waves 28(1), 13–23 (2007)

102. Pozar, D.M.: A review of aperture coupled microstrip antenna: history, operation, development, and applications. In: Electrical and Computer Engineering. University of Massachusetts at Amherst, Amherst (1996), http://www.ecs.umass.edu/ece/pozar/aperture.pdf (last accessed on September 08, 2011)

103. Wong, K.L., Wu, J.Y.: Single fed small circularly polarized square microstrip antenna. Electron. Letters 33(22), 1833–1834 (1997)

104. Garg, R., Bharti, P., Bahl, I., Ittipiboon, A.: Microstrip antenna design handbook. Aartech House, London (2001)

105. Balanis, C.A.: Antenna theory - analysis and design, 2nd edn. John Wiley & Sons, New York (1997)

106. Bancroft, R.: Microstrip and printed antenna design. Prentice-Hall of India, New Delhi (2006)

107. Nasimuddin, Esselle, K.P., Verma, A.K.: Improving the axial-ratio bandwidth of circularly polarized stacked microstrip antennas and enhancing their gain with short horns. In: Proceedings of IEEE Antennas and Propagation Society (AP-S) International Symposium 2006, Albuquerque, NM, USA, pp. 1545–1548 (July 9-14, 2006)

108. Gunn, S.R.: Support vector machine Matlab toolbox (1998), http://www.isis.ecs.soton.ac.uk/resources/svminfo/ (last accessed on November 2008)

109. Nasimuddin, Esselle, K.P., Verma, A.K.: Fast and accurate model for circular microstrip antennas on suspended and composite substrates. IEEE Transactions on Antennas and Propagation 53(9), 3097–3100 (2005)

110. Kartikeyan, M.V., Borie, E., Thumm, M.K.A.: Gyrotrons - high power microwave/millimeter wave technology. Springer, Berlin (2004)

111. Grudiev, A., Schünemann, K.: Nonstationary behavior of the gyrotron backward-wave oscillator. IEEE Transactions on Plasma Science 30(3), 851–858 (2002)

112. Schünemann, K., Serebryannikov, A., Vavriv, D.: Analysis of the complex natural frequency spectrum of the azimuthally periodic coaxial cavity. Microwave and Optical Technology Letters 17(5), 308–313 (1998)

113. Chauhan, N.C., Mittal, A., Wagner, D., Kartikeyan, M.V., Thumm, M.K.: Design and optimization of nonlinear tapers using particle swarm optimization. Journal of Infrared, Millimeter and Terahertz Waves 29(8), 792–798 (2008)

114. Chauhan, N.C., Kartikeyan, M.V., Mittal, A.: Design of critical output components of high power gyrotrons using particle swarm optimization. In: Proceedings of National Symposium on Vacuum Electronic Devices & Applications (VEDA-2009), pp. Gyro3.1-Gyro3.3. Institute of Technology, Banaras Hindu University, Varanasi (January 8-10, 2009)

115. Chauhan, N.C., Kamakshi, S., Kumar, A., Wagner, D., Kartikeyan, M.V., Mittal, A., Thumm, M.: Design and optimization of nonlinear tapers for high power gyrotrons. In: National Symposium on Plasma (PLASMA-2007). Institute of Plasma Research, Gandhinagar (2007)

116. Bogdashov, A.A., Rodin, Y.V.: Mode converter synthesis by the particle swarm optimization. International Journal of Infrared Millimeter Waves 28, 627–638 (2007)

117. Chauhan, N.C., Vasista, B.R., Sandeep Reddy, M., Kartikeyan, M.V., Mittal, A.: Modified Bacterial Foraging Optimization and its Application for the Design of a Nonlinear Taper. In: Proceedings of International Conference on Advanced Computing Technologies (ICACT-2008), Hyderabad, India, pp. 23–28 (2008)

118. Collin, R.E.: Foundations for microwave engineering, 2nd edn. McGraw-Hill, New York (1992)

119. Dwari, S., Chakraborty, A., Sanyal, S.: Analysis of linear tapered waveguide by two approaches. Progress In Electromagnetics Research (PIER) 64, 219–238 (2006)

120. Klopfenstein, R.W.: A transmission line taper of improved design. Proceedings of the Institute of Radio Engineers 44(1), 31–35 (1956)

121. Hecken, R.P.: A near-optimum matching section without discontinuities. IEEE Transactions on Microwave Theory and Techniques 20(11), 734–739 (1972)

122. Flügel, H., Kühn, E.: Computer-aided analysis and design of circular waveguide tapers. IEEE Transactions on Microwave Theory and Techniques 36(2), 332–336 (1988)

123. Lawson, W.G.: Theoretical evaluation of non-linear tapers for a high power gyrotrons. IEEE Transactions on Microwave Theory and Techniques 38(11), 1617–1622 (1990)

124. Tripathy, M., Mishra, S.: Bacterial foraging based solution to optimize both real power loss and voltage stability limit. IEEE Transactions on Power Systems 22(1), 240–408 (2007)

125. Wagner, D., Thumm, M., Gantenbein, G., Kasperek, W., Idehara, T.: Analysis of a complete gyrotrons oscillator using the scattering matrix description. International Journal of Infrared and Millimeter Waves 19(2), 185–194 (1998)

126. Wagner, D., Kasparek, W., Thumm, M.: Optimized nonlinear cavity and up-taper for gyrotron FU V. In: Proceedings of the 11th Joint Russian-German Meeting on ECRH and Gyrotrons, Karlsruhe, Stuttgart, Garching, pp. 99–110 (June 23-29, 1999)

127. Eichmeier, J.A., Thumm, M. (eds.): Vacuum electronics - components and devices, 1st edn. Springer, Heidelberg (2008)

128. Thumm, M.: Development of output windows for high power long pulse gyrotron and EC wave applications. International Journal of Infrared and Millimeter Waves 19(1), 3–14 (1998)

129. Yang, X., Wagner, D., Piosczyk, B., Koppenberg, K., Borie, E., Heidinger, R., Leuterer, F., Dammertz, G., Thumm, M.: Analysis of transmission characteristics for single and double disc windows. International Journal of Infrared Millimeter Waves 24(5), 619–628 (2003)

130. Jöstingmeier, A., Dohlus, M., Holtkamp, N.: Systematic design of an S-band pillbox type RF window. In: XIX International Linear Accelerator Conference (LINAC-1998), Chicago, USA, August 23-28, pp. 249–251 (1998)

131. Yang, X., Borie, E., Dammertz, G., Heidinger, R., Koppenburg, K., Leuterer, F., Piosczyk, B., Wagner, D., Thumm, M.: The influence of window parameters on the transmission characteristics of millimeter waves. International Journal of Infrared Millimeter Waves 24(11), 1805–1813 (2003)

132. Chauhan, N.C., Kartikeyan, M.V., Mittal, A.: CAD of RF windows using multi-objective particle swarm optimization. IEEE Transactions on Plasma Science 37(6), 1104–1109 (2009)

133. Chauhan, N.C., Kartikeyan, M.V., Mittal, A.: Design of RF window using multi-objective particle swarm optimization. In: Proceedings of International Conference on Recent Advances in Microwave Theory and Applications (MICROWAVE-2008), November 21-24, pp. 34–37. University of Rajasthan, Jaipur (2008)

134. Jena, R., Sharma, G., Mahanti, P.K.: A multi-objective optimization model for SoC design space exploration at system level. Journal of Advanced Manufacturing Technology 1(1), 101–111 (2007)

135. Gies, D., Rahmat-Samii, Y.: Vector Evaluated Particle Swarm Optimization (VEPSO): optimization of a radiometer array antenna. In: IEEE Antenna and Propagation Society International Symposium, Monterey, CA, USA, June 20-26, pp. 2297–2300 (2004)

136. Raquel, C.R., Naval, P.C.: An effective use of crowding distance in multi-objective particle swarm optimization. In: Proceedings of Genetic and Evolutionary Computation Conference (GECCO 2005), Washington DC, pp. 257–264 (June 2005)

137. Kartikeyan, M.V., Borie, E., Piosczyk, B., Lamba, O.S., Singh, V.V.P., Möbius, A., Bandopadhyay, H.N., Thumm, M.: Conceptual design of a 42 GHz, 200 kW gyrotron operating in the $TE_{5,2}$ mode. International Journal of Electronics 87(6), 709–723 (2000)

138. Singh, V.V.P., Chander, S., Joshi, L.M., Sinha, A.K., Bandopadhyay, H.N.: On the electrical design of pill-box type high power microwave window. Journal of IETE 39(6), 351–359 (1993)

139. Nickel, H. -U.: Hochfrequenztechnische aspekte zur entwicklung riick-wirkungsarmer ausgangsfenster für millimeterwellen- gyrotrons hoher leistung. PhD Thesis. University of Karlsruhe, Germany (1995)
140. Cascade Scattering Matrix Code, v3.0, Calabazas Creek Research, Inc., (2000)
141. Deb, K.: Multi-objective optimization using evolutionary algorithms. Wiley, Chichester (2001)
142. Chauhan, N.C., Aggarwal, D., Banga, R., Mittal, A., Kartikeyan, M.V.: Parallelization of particle swarm optimization and its implementation on scalable multi-core architecture. In: IEEE International Advance Computing Conference (IACC-2009), March 6-7, Patiala, India (2009)